Functional Synthetic Polymers

Scrivener Publishing
100 Cummings Center, Suite 541J
Beverly, MA 01915-6106

Publishers at Scrivener
Martin Scrivener (martin@scrivenerpublishing.com)
Phillip Carmical (pcarmical@scrivenerpublishing.com)

Functional Synthetic Polymers

Johannes Karl Fink

Scrivener Publishing

WILEY

This edition first published 2019 by John Wiley & Sons, Inc., 111 River Street, Hoboken, NJ 07030, USA and Scrivener Publishing LLC, 100 Cummings Center, Suite 541J, Beverly, MA 01915, USA

For more information about Scrivener publications please visit www.scrivenerpublishing.com.

Wiley Global Headquarters
111 River Street, Hoboken, NJ 07030, USA

For details of our global editorial offices, customer services, and more information about Wiley products visit us at www.wiley.com.

Library of Congress Cataloging-in-Publication Data
ISBN 978-1-119-59202-0

Cover images: Pixabay.Com
Cover design by: Russell Richardson

Set in size of 11pt and Minion Pro by Exeter Premedia Services Private Ltd., Chennai, India

10 9 8 7 6 5 4 3 2 1

Contents

Preface

The scientific literature with respect to functional synthetic polymers is collected in this monograph. The text focuses on the basic issues and also the literature of the past decades. The book provides a broad overview of the synthesis procedures for functional synthetic polymers and the materials used therein.

In addition to basic issues concerning functionalized polymers, particular emphasis is given to the principles of functionalization, basic functional groups, and surface functionalization. Also, fields of special application, such as electrical applications, water cleaning methods, and medical and pharmaceutical applications, are reviewed.

Beyond educating students of polymer chemistry, this book will be of importance to chemists and other scientists in specialty fields, such as electronics, medicine and pharmacology, interested in expanding their knowledge about topics concerning the issues in this field.

Among the special issues addressed in the text are: Surface functionalization supramolecular polymers, shape-memory polymers, foldable polymers, functionalized biopolymers, supercapacitors, photovoltaics, lithography, cleaning methods, such as recovery of gold ions olefin/paraffin, separation by polymeric membranes, ultrafiltration membranes, and other related topics.

How to Use This Book

Utmost care has been taken to present reliable data. However, because of the vast variety of material presented herein, it is not possible to include detailed information on all aspects of the topic, and it is recommended that the reader study the original literature for more complete information.

Index

There are three indices: an index of acronyms, an index of chemicals, and a general index. In the index of chemicals, compounds that occur extensively are not included at every occurrence, but rather when they appear in an important context. When a compound is found in a figure, the entry is marked in boldface letters in the chemical index.

Acknowledgements

I am indebted to our university librarians, Dr. Christian Hasenhüttl, Dr. Johann Delanoy, Gerlinde Iby, Franz Jurek, Margit Keshmiri, Friedrich Scheer, Christian Slamenik, Renate Tschabuschnig, and Elisabeth Groß for their support in literature acquisition. In addition, many thanks to the head of my department, Professor Wolfgang Kern, for his interest and permission to prepare this text.

I also want to express my gratitude to all the scientists who have carefully published their results concerning the topics dealt with herein. This book could not have been otherwise compiled.

Last, but not least, I want to thank the publisher, Martin Scrivener, for his abiding interest and help in the preparation of the text. In addition, my thanks go to Jean Markovic, who made the final copyedit with utmost care.

Johannes Fink
Leoben, 7th February 2019

1

Basic Issues of Functionalized Polymers

In organic chemistry, functionality is often used as a synonym for a functional group. Functionalization means the introduction of functional groups. According to IUPAC (1), the functionality of a monomer is defined as the number of bonds that a repeating unit of a monomer forms in a polymer with other monomers. Thus, in the case of a functionality of $f = 2$, a linear polymer is formed by polymerizing a thermoplastic material.

Monomers with a functionality of > 3 lead to a branching point, which can result in crosslinked polymers, i.e., a thermosetting polymer (2).

Functional polymers are also sometimes called smart polymers (3). Here, the basic issues of functionalized polymers are discussed.

Functional polymers are polymers with advanced optic and/or electronic properties. The advantages of functional polymers include their low cost, the ease in which they can be processed and a range of attractive mechanical characteristics for functional organic molecules.

These properties can be adjusted whilst material usage is kept low, consequently opening interesting environmental perspectives. Polymer-bound substances can spread their activity without endangering people or the environment. Examples of functional polymers are (4):

1. Semiconducting conjugated polymers.

2. Stimuli-responsive polymers,
3. Thermally responsive polymers,
4. pH-sensitive polymers,
5. Toner materials,
6. Supercapacitors,
7. Ion exchange resins,
8. Biomimetic materials, and
9. Supramolecular metallopolymers.

Functional polymers are macromolecules to which chemically bound functional groups are attached which can be utilized as reagents, catalysts, protecting groups, and others. Functional polymers are low cost, easy to process and have a range of attractive mechanical characteristics for functional organic molecules.

The polymer support can be either a linear species, which is soluble, or a crosslinked species which is insoluble. A polymer that can be used as support should have significant mechanical stability under the reaction conditions. Such properties of the support play an important role in the functionalization reactions of polymers. So, the polymer properties can be modified either by chemical reactions on pendant groups or by changing the physical nature of the polymers. Special uses of functional polymers are shown in Table 1.1.

Table 1.1 Uses of functional polymers (5).

Field of Application	Use
Analytical chemistry	Polymers as stationary-phase (chromatography/extraction)
Catalysis engineering	Polymers as a catalyst
Medicine, agriculture, washing agents	Controlled release from polymer matrices, design and synthesis of functional polymers, polymer-bound dyes, reactive and functional polymers
Polymer modification	Surface and functional coatings

There are monographs dealing with functional polymers (6–14).

A comprehensive and authoritative overview of functional polymers and polymeric materials has been presented (14). This ranges from their synthesis and characterization, to their properties, actual applications and future perspectives.

Functional polymers and smart polymeric materials play a decisive role in new innovations in all areas where new materials are needed. Optoelectronics, catalysis, biomaterials, medicine, building materials, water treatment, coatings, and many more applications rely on functional polymers.

Functional polymers are polymers that respond to different stimuli or changes in the environment. The types of polymers, including temperature-, pH-, photo-, and enzyme-responsive polymers, have been assessed (10). These issues include shape-memory polymers, smart polymer hydrogels, and self-healing polymer systems.

Applications of functional polymers include smart instructive polymer substrates for tissue engineering, smart polymer nanocarriers for drug delivery, the use of smart polymers in medical devices for minimally invasive surgery, diagnosis, and other applications, and smart polymers for bioseparation and other biotechnology applications. Functional polymers are also used for textile and packaging applications, and for optical data storage.

Adaptive polymers are those which are responsive to different stimuli, namely physical, mechanical, chemical and biological stimuli, with a controlled and/or predicable behavior. They can be used in textiles, skin care, medicine and other related areas. Some versatile functional polymers, such as chitosan, cylodextrin and dendrimer, and hyperbranched polymers have also been reviewed (8).

Functional polymers are also important materials for coatings (11). For example, superhydrophobic surfaces can be produced.

Also, functional biopolymers have been reviewed (15). A comprehensive overview of the synthesis, properties and biomedical applications of functional biopolymers has been presented. A lot of topics are covered, such as synthetic biopolymers, blood-compatible polymers, ophthalmic polymers and stimuli-responsive polymers. An up-to-date review of cell encapsulation strategies and cell surface and tissue engineering has also been included in this work.

1.0.1 Standards

Actually, standards specifically designed for functional polymers are rare. No standard with the term *functional polymer* in the title could be found. However, in the scientific literature, in the context of functional polymers, some standards for measuring the properties of these polymers have been mentioned. These standards are collected in Table 1.2.

The *Functional Polymers Group* of the National Institute of Standards and Technology in Gaithersburg, MD, develops directives, measurement methods, data, standards, and science for the functional properties (e.g., electronic, ion transport) of polymeric materials within functional devices and applications in forms that include thin films, interfaces, nanostructures, and membranes (27). The projects and programs of this group are summarized in the following sections.

1.0.1.1 Polymers for Next-Generation Lithography

Here, measurements were developed and applied with high-spatial and chemically specific resolution to elucidate the properties and process kinetics of critical materials at nanometer scales that are needed to advance next-generation photolithography, including both the 193 *nm* (deep ultraviolet) and 13.5 *nm* (extreme ultraviolet) lithography platforms.

This provides the foundation for a rational design of materials and processing strategies for the fabrication of sub-32 *nm* structures. The measurement platform integrates specular and off-specular X-ray and neutron reflectivity, near-edge X-ray absorption fine structure spectroscopy, quartz crystal microbalance, solid-state nuclear magnetic resonance, polarization-modulation infrared reflectance absorption spectroscopy, and infrared variable angle spectroscopic ellipsometry. These measurements relate the fundamentals of polymer interfaces to high-resolution lithographic patterning.

1.0.1.2 Templated Assembly of Block Copolymer Films

Block copolymers are materials that naturally self-assemble into monodisperse, chemically distinct domains, however, their placement and orientation are difficult to control. Recent work has demonstrated a capability to utilize these domains as masks for pattern transfer and as functional materials for membranes.

Viable nanomanufacturing of templated block copolymers will require a capability to control the orientation and the line edge

Table 1.2 Standards in the context of functional polymers.

Number	Title	Reference
ASTM D3643-15	Standard Test Method for Acid Number of Certain Alkali-Soluble Resins	(16)
ASTM D790-17	Standard Test Methods for Flexural Properties of Unreinforced and Reinforced Plastics and Electrical Insulating Materials	(17)
ASTM D785-08	Standard Test Method for Rockwell Hardness of Plastics and Electrical Insulating Materials	(18)
ASTM D638-14	Standard Test Method for Tensile Properties of Plastics	(19)
ASTM D570-98	Standard Test Method for Water Absorption of Plastics	(20)
ASTM F813-07	Standard Practice for Direct Contact Cell Culture Evaluation of Materials for Medical Devices	(21)
ASTM F619-14	Standard Practice for Extraction of Medical Plastics	(22)
ASTM D5229	Standard Test Method for Moisture Absorption Properties and Equilibrium Conditioning of Polymer Matrix Composite Materials	(23)
ASTM D2872-12e1	Standard Test Method for Effect of Heat and Air on a Moving Film of Asphalt (Rolling Thin-Film Oven Test)	(24)
ISO 17736	Workplace air quality – Determination of isocyanate in air using a double-filter sampling device and analysis by high pressure liquid chromatography	(25)
ISO 8986-1	Plastics – Polybutene-1 (PB-1) moulding and extrusion materials – Part 1: Designation system and basis for specifications	(26)

roughness of trillions of structures to within a single nanometer. However, there are no existing platforms that meet this need. Small-angle X-ray and neutron scattering techniques were developed to measure the orientation, orientation distribution, and pattern quality of block copolymers ordered on chemical and physical templates over large areas with high precision.

These parameters are the key to developing effective nanomanufacturing strategies. A methodology was developed to create large area block copolymer films with a highly preferred orientation. These samples are then suitable for characterization by small-angle scattering to determine an orientation distribution indicative of pattern quality, including line edge roughness.

This capability provides powerful data on the relationship between materials, processing, template design, and the expected pattern quality.

1.0.1.3 Sustainable Composites

The objective of this project is the development of tools to measure the fundamental structure-processing and structure-property relations associated with sustainable polymer composites.

This will be accomplished by using interface characterization methods, such as Förster resonance energy transfer, nuclear magnetic resonance spectroscopy and Raman spectroscopy, to quantify the effects of the complex interactions and high degree of chemical functionality characteristic of the interface of these materials.

These methods will be coupled with mechanical property measurements to characterize the effect of mechanical degradation, aging, and hydrothermal effects on interface structure and integrity in sustainable/conventional composite blends and biobased polymer nanocomposites.

1.0.1.4 Energy Storage and Delivery

The structure and dynamics of important classes of polymer electrolyte membrane materials should be elucidated, including emerging systems like block copolymers, polymer blends, and candidate materials.

Advanced methods were developed that illuminate the relationship between the molecular architecture and the resulting nanostructure of a polymer electrolyte membrane material, information which is key to understanding the membrane performance, including durability and proton conductivity. The measurement methods should provide the precise 3D nanoscale morphological information that is missing from conventional analyses of these materials.

Small-angle neutron scattering techniques are developed that characterize the structure of hydrated, nanostructured membranes based on block copolymers and complex polymer blends.

In addition, quasielastic neutron scattering methods are tested for the analysis of polymer and water dynamics in both dry and hydrated membrane materials. Quasielastic neutron scattering was used to measure the correlations between counter-ion dynamics and bulk mechanical relaxations in alkyl ammonium neutralized membranes. These measurements have provided a physical description of how the network structure contributes to the ionic conductivity properties of the membrane.

1.0.1.5 Polymer Membranes

Membranes and membrane technology are key to water and energy security. Polymer-based membranes already play a significant role in fields such as impact ballistic testing of polymer films.

The *Membranes for Clean Water* project provides measurement solutions that probe the surface and internal structure of polymer membranes used in water purification, and correlate that structure to the transport of water and other species through the membrane. The methods are focused on elucidating the role of roughness, surface charge, surface chemistry, crosslink density and monomer chemistry on the interfacial and internal dynamics of the membrane. With this knowledge, the industry will better understand the performance of these membrane materials, as well as identify essential features to enable next-generation, energy-efficient, high-flux membranes.

1.0.1.6 *Mechanics of Polymers and Interfaces*

Polymer films and coatings play a central role in high impact mitigation applications ranging from helmets, to body armor, to aircraft and spacecraft.

Although there are very many energy dissipating mechanisms available for polymeric materials, the specific mechanism for high impact scenarios depends on the polymer structure and is not well understood.

Specifically, the mechanical testing infrastructure is developed in order to establish relationships between the polymer structure and physics that underpin the mechanical performance of engineered structures, composite interfaces, thin films and membranes. The investigated methods are laser-induced projectile impact testing, wrinkling-cracking, cavitation rheology, contact adhesion testing, poromechanical relaxation indentation, and neutron scattering.

Additionally, by coupling the results from these experimental measurements with theory, simulation, and modeling, the computational design of energy mitigating materials is facilitated.

References

1. A.D. McNaught and A. Wilkinson, eds., *IUPAC. Compendium of Chemical Terminology*, the Gold Book, Blackwell Scientific Publications, Oxford, 2nd edition, 1997.
2. Wikipedia contributors, Functionality (chemistry) — Wikipedia, the free encyclopedia, https://en.wikipedia.org/w/index.php?title=Functionality_(chemistry)&oldid=804812838, 2017. [Online; accessed 27-October-2018].
3. Wikipedia contributors, Smart polymer — Wikipedia, the free encyclopedia, https://en.wikipedia.org/w/index.php?title=Smart_polymer&oldid=861339882, 2018. [Online; accessed 1-October-2018].
4. Wikipedia Contributors, Functional polymers — Wikipedia, the free encyclopedia, https://en.wikipedia.org/w/index.php?title=Functional_polymers&oldid=859144249, 2018. [Online; accessed 28-September-2018].
5. OMICS International, Functional polymer and its applications, electronic: https://www.omicsonline.org/conferences-list/functional-polymer-and-its-applications, 2018.
6. R. Arshady and A. Guyot, eds., *Functional Polymer Colloids & Microparticles*, Citus Reference Series, Citus, London, 2002.

7. M.R. Aguilar, C. Elvira, A. Gallardo, B. Vázquez, and J. Román, *Smart polymers and their applications as biomaterials* in N. Ashammakhi, R. Reis, and E. Chiellini, eds., *Topics in Tissue Engineering*, Vol. 3, chapter 6, pp. 1–27. Woodhead Publishing, an imprint of Elsevier Science, Sawston, UK, 2007.
8. J. Hu, *Adaptive and Functional Polymers, Textiles and their Applications*, Imperial College Press, London, 2011.
9. V. Mittal, ed., *Functional Polymer Blends: Synthesis, Properties, and Performances*, CRC Press, Boca Raton, 2012.
10. M. Aguilar and J. San Román, eds., *Smart Polymers and their Applications*, Woodhead Publishing, an imprint of Elsevier Science, Sawston, UK, 2014.
11. L. Wu and J. Baghdachi, eds., *Functional Polymer Coatings: Principles, Methods and Applications*, Wiley Series on Polymer Engineering and Technology, John Wiley & Sons Inc, Hoboken, New Jersey, 2015.
12. Y. Lvov, B. Guo, and R.F. Fakhrullin, eds., *Functional Polymer Composites with Nanoclays*, Royal Soc. of Chemistry, Cambridge, 2017.
13. R. Shunmugam, ed., *Functional Polymers: Design, Synthesis, and Applications*, Apple Academic Press, Oakville, ON, Canada, Waretown, NJ, USA, 2017.
14. M.A.J. Mazumder, H. Sheardown, and A. Al-Ahmed, eds., *Functional Polymers*, Polymers and Polymeric Composites, Springer International Publishing, Cham, Switzerland, 2019.
15. V.K. Thakur and M.K. Thakur, eds., *Functional Biopolymers*, Polymers and Polymeric Composites, Springer International Publishing, Cham, Switzerland, 2018.
16. Subcommittee: D21.02, Standard test method for acid number of certain alkali-soluble resins, ASTM Standard D3643-15, ASTM International, West Conshohocken, PA, 2015.
17. Subcommittee: D20.10, Standard test methods for flexural properties of unreinforced and reinforced plastics and electrical insulating materials, ASTM Standard D790-17, ASTM International, West Conshohocken, PA, 2017.
18. Subcommittee: D20.10, Standard test method for Rockwell hardness of plastics and electrical insulating materials, ASTM Standard D785-08, ASTM International, West Conshohocken, PA, 2015.
19. Subcommittee: D20.10, Standard test method for tensile properties of plastics, ASTM Standard D638-14, ASTM International, West Conshohocken, PA, 2014.
20. Subcommittee: D20.50, Standard test method for water absorption of plastics, ASTM Standard D570-98, ASTM International, West Conshohocken, PA, 2018.

21. Subcommittee: F04.16, Standard practice for direct contact cell culture evaluation of materials for medical devices, ASTM Standard F813-07, ASTM International, West Conshohocken, PA, 2012.
22. Subcommittee: F04.16, Standard practice for extraction of medical plastics, ASTM Standard F619-14, ASTM International, West Conshohocken, PA, 2014.
23. Subcommittee: D30.04, Standard test method for moisture absorption properties and equilibrium conditioning of polymer matrix composite materials, ASTM Standard D5229/D5229M-14, ASTM International, West Conshohocken, PA, 2014.
24. Subcommittee: D04.46, Standard test method for effect of heat and air on a moving film of asphalt (rolling thin-film oven test), ASTM Standard D2872-12e1, ASTM International, West Conshohocken, PA, 2012.
25. Technical Committee: ISO, Workplace air quality – determination of isocyanate in air using a double-filter sampling device and analysis by high pressure liquid chromatography, ISO Standard 17736, International Organization for Standardization, Geneva, Switzerland, 2010.
26. Technical Committee: ISO, Plastics – polybutene-1 (pb-1) moulding and extrusion materials – part 1: Designation system and basis for specifications, ISO Standard 8986-1, International Organization for Standardization, Geneva, Switzerland, 2009.
27. C. Soles, Functional polymers group, electronic: https://www.nist.gov/mml/materials-science-and-engineering-division/functional-polymers-group, 2018.

2

Methods and Principles of Functionalization

The synthesis of functional polymers of polar and nonpolar monomers by living and/or controlled polymerization has been detailed in a monograph (1).

2.1 Analysis and Characterization

2.1.1 Multidimensional Mass Spectrometry

Multidimensional mass spectroscopy (MS) interfaces a suitable ionization technique and mass analysis with fragmentation by tandem mass spectrometry and an orthogonal online separation method (2).

Separation choices include liquid chromatography (LC) and ion-mobility spectrometry, in which separation takes place preionization in the solution state or post-ionization in the gas phase, respectively. The mass spectrometry step provides information on the elemental composition while tandem mass spectrometry assesses differences in the bond stabilities of a polymer, yielding connectivity and sequence information. The conditions of LC can be tuned to separate by polarity, end-group functionality, or hydrodynamic volume, whereas ion-mobility spectrometry adds selectivity by macromolecular shape and architecture.

There has been a discussion on, how selected combinations of MS, tandem mass spectrometry, LC, and ion-mobility spectrometry can be applied, together with the appropriate ionization method, to determine the constituents, structures, end groups, sequences, and architectures of a wide variety of homopolymers and copolymeric materials (2).

The m/z information available through MS measurement is often sufficient to deduce a polymer's compositional heterogeneity as well as its total chain end-group and other functionality distributions (3–6).

Several examples of spectra concerning functional groups in polymers have been presented and discussed (2). The MS characteristics of certain polymers depend on the fragmentation energetics of their blocks, which can differ considerably depending on the functional groups and the connectivity of their repeat units.

2.1.2 *Analysis of Phosphorus-End Capped Functional Polymers*

A hyphenated method based on thermal gravimetric, gas chromatographic and mass spectrometric analysis has been proposed for the quantitative analysis of diethyl phosphate-end capped polymers with varying chain length, in bulk and as ultrathin film (7).

This method was revealed to be a powerful and affordable tool in the determination of the amount of phosphorous in polymeric samples. It also allows the determination of the number of repeating units per chain end.

Here, in comparison to common techniques, no sample pretreatment is needed. Consequently, polymeric films supported on inert substrates can be directly analyzed. The quantitative calibration obtained for bulk materials was also demonstrated to be accurate for films of different thickness. In this frame, the method was revealed to be highly sensitive, thus allowing for the determination of phosphorous even in ultrathin films with a thickness of a few tens of nanometers (7).

2.2 Functional Groups

The interaction of polymers with their environment depends largely on the functional groups that they are carrying (8). Interfaces between different polymers or between polymers and other surfaces can be strengthened through the design of molecular interactions, such as hydrogen bonding, and through the control of the polymer architecture. The placement of functional groups at the ends of polymer chains or in well-defined segments can determine the ultimate properties. Three-dimensional synthetic polymers, such as dendrimers, can be fashioned to encapsulate reactive sites or provide highly controlled surfaces and interfaces.

2.3 Stille Polycondensation

The Stille reaction, also known as Migita-Kosugi-Stille coupling, is a chemical reaction that is widely used in organic synthesis. It involves the coupling of an organotin compound with a variety of organic electrophiles via a palladium-catalyzed coupling reaction (9). The basic mechanism runs as follows:

$$R_1{-}Sn(Alkyl)_3 + R_2{-}X \rightarrow R_1{-}R_2 + X{-}Sn(Alkyl)_3 \quad . \qquad (2.1)$$

Also, the Stille reaction has been used for the synthesis of a variety of polymers (10–12). Stille polycondensation is a very versatile synthetic method for conjugated polymers because of its efficiency, mild reaction conditions and tolerance to a wide scope of functional groups (13).

Many aspects of Stille polycondensation have been detailed in a monograph (13):

1. Its history and investigation of the mechanism and reaction conditions,
2. Its application in preparing numerous kinds of functional materials, including nonlinear optical polymers, organic photovoltaic polymers, organic field effect transistor polymers, organic light-emitting diode polymers, bio and chemical sensors, and
3. New developments and future issues.

2.4 Light-Mediated Atom Transfer Radical Polymerization

A highly efficient photomediated atom transfer radical polymerization (ATRP) protocol has been reported for semi-fluorinated acrylates and methacrylates (14).

The use of a commercially available solvent, i.e., 2-trifluoromethyl-2-propanol, optimally balances monomer, polymer, and catalyst solubility while eliminating transesterification as a detrimental side reaction. 2-Trifluoromethyl-2-propanol is shown in Figure 2.1

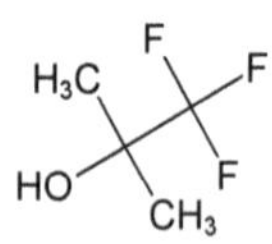

Figure 2.1 2-Trifluoromethyl-2-propanol.

In the presence of UV irradiation and *ppm* concentrations of copper(II) bromide and Me_6-tris(2-aminoethyl amine), semi-fluorinated monomers with side chains containing between three and 21 fluorine atoms can readily polymerize under controlled conditions.

The resulting polymers exhibit narrow molar mass distributions (M_w/M_n = 1.1) and a high end group fidelity, even at conversions greater than 95%.

This level of control permits the *in-situ* generation of chain-end functional homopolymers and diblock copolymers, providing facile access to semi-fluorinated macromolecules using a single methodology with unprecedented monomer scope. The results disclosed herein should create opportunities across a variety of fields that exploit fluorine-containing polymers for tailored bulk, interfacial, and solution properties.

2.5 Catalytic Insertion Polymerization

PhS- and PhNH-functionalized dienes could be copolymerized efficiently with butadiene to stereoregular copolymers by [(mesitylene)Ni(allyl)] [$BArF_4$] (Ni-1) (15). The overall polymerization rates and comonomer incorporations depend strongly on the linker length between the diene moiety and functional group, in, e.g., $PhS-(CH_2)_xC(-CH_2)-CH-CH_2$ ($x = 3-7$), in particular for certain linker lengths high comonomer reactivity ratios stand out.

This effect is related to a favorable binding of the comonomer to the active site comprising coordination of its functional group, which significantly enhances the comonomer incorporation in the growing polymer chain (15).

2.6 C–H Functionalization

Homogeneous catalysis of C–H activation and functionalization reactions by transition metal complexes go back to Fenton in 1894 (16). The modern development of this topic is known as Shilov's platinum chemistry (17,18).

The late stage functionalization (19) of complex organic molecules provides a potential way to simplify the synthetic strategies as well as permits a wider variety of final products to be made from the same late stage intermediate (20).

Electrochemical transition metal catalysis is a powerful strategy for organic synthesis because it obviates the use of stoichiometric chemical oxidants and reductants (21). C–H bond functionalization offers a variety of useful conversions of simple and ubiquitous organic molecules into diverse functional groups in a single synthetic operation.

The recent progress in merging electrochemistry with transition metal-catalyzed C–H functionalization has been reviewed, specifically C–C, C–X (halogen), C–O, C–P, and C–N bond formation (21).

Also, the decarboxylative C–H bond functionalization in the synthesis of various organic compounds, such as styrenes, chalcones, biaryls, and heterocycles, has been reviewed (22).

Direct transformations of C–H bonds have long been known from reactions in organic chemistry as unselective oxidation processes. However, more recent efforts have produced refinements that offer high selectivity, and it is for these processes that the term C–H functionalization is currently used. The C–H bond has been called the unfunctional group (23).

Reaction types that can selectively functionalize C–H bonds are of intense interest because they offer new strategic approaches for synthesis (24). A very promising C–H functionalization method involves the insertion of metal carbenes and nitrenes into C–H bonds. This area has experienced considerable growth in the past decade, particularly in the area of enantioselective intermolecular reactions.

Several facets of these kinds of C–H functionalization reactions have been discussed (24). Also, a perspective is provided on how this methodology has affected the synthesis of complex natural products and potential pharmaceutical agents.

The C–H functionalization via a metal carbenoid approach typically uses a high-energy diazo compound and the loss of nitrogen provides the driving force for the energetically unfavorable formation of the carbenoid. The highly reactive carbenoid species then inserts into a C–H bond to form the C–H activation product and liberates the metal catalyst for another cycle.

The substituents attached to the carbenoid help to modulate its reactivity. The presence of both an electron-donating group and an electron-withdrawing group is necessary to reduce carbene dimerization pathways and increase selectivity for intermolecular reactions. During the C–H activation event, a partial positive charge buildup occurs at the carbon undergoing C–H functionalization. Sites adjacent to functionality that can stabilize this polarization are considered to be electronically activated towards carbenoid reactions (24).

The synthesis of [5,6]-bicyclic heterocycles has been reported. Here, a ring-junction nitrogen is produced via a Rh^{3+} catalyzed C–H functionalization of alkenyl azoles (25). Several reaction types were applied to alkenyl imidazoles, pyrazoles, and triazoles to provide products with nitrogen incorporated at different sites. Alkyne and diazoketone coupling partners give azolopyridines with various substitution patterns. In addition, 1,4,2-dioxazolone coupling partners yield azolopyrimidines. The synthesis of azolopyridines and azolopyrimidines is shown in Figure 2.2.

2.6.1 *Enantioselective Resin*

An enantioselective *l*-histidine imprinted polymeric resin was fabricated and evaluated for the enantiomeric resolution of histidine

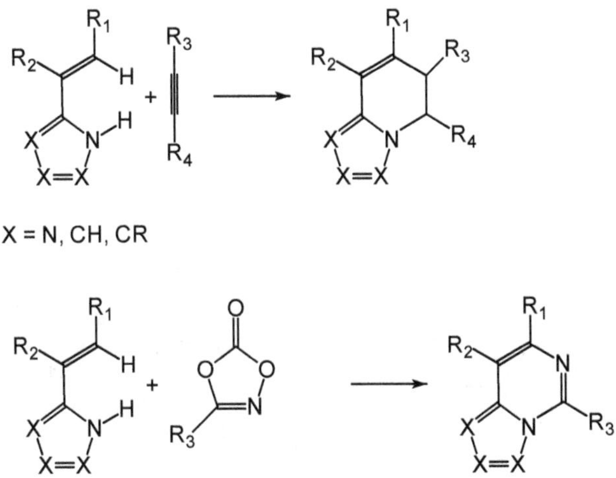

Figure 2.2 Synthesis of azolopyridines and azolopyrimidines (25).

racemate. First, the polymerizable chiral salicyloyl-*l*-histidine amide was synthesized (26). The synthesis is shown in Figure 2.3.

This compound was then anchored onto a polymeric resin network via the condensation polymerization of resorcinol and formaldehyde.

The *l*-histidine template molecules were then extracted out of the resin texture via alkaline hydrolysis of the amide bond using sodium hydroxide (26).

The complete extraction of the template *l*-histidine molecules was assured using energy-dispersive X-ray spectroscopy, which indicated the absence of nitrogen upon alkaline treatment of the synthesized *l*-histidine containing resin.

Selective adsorption experiments indicated that the maximum adsorption was achieved at pH 8 and followed the pseudo-second-order kinetic model with extracted amounts of 165±1 and 90±1 $mg\,g^{-1}$ with respect to *l*-histidine and *d*-histidine, respectively.

A Langmuir model displayed the best fit with the experimentally obtained isotherm data. The maximum adsorption capacities were 195±1 and 102±1 $mg\,g^{-1}$ with respect to *l*-histidine and *d*-histidine, respectively. The enantiomeric resolution of a d/*l*-histidine racemate was also carried out utilizing a column backed with the imprinted

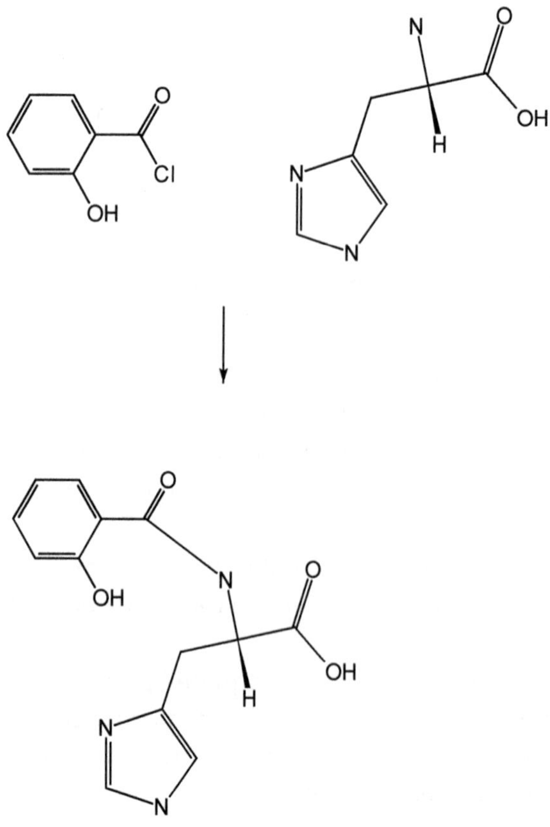

Figure 2.3 Synthesis of salicyloyl-*l*-histidine amide (26).

resin and the outlet collected solution displayed an optical activity related to 36% *d*-histidine enantiomeric excess (26).

2.7 Surface Functionalization

2.7.1 *Pigmented Coating Composition*

The hiding efficiency in pigmented paints can be improved by using an itaconic acid, c.f. Figure 2.4, functionalized latex binder in combination with a solution of a polymer containing sulfonic acid groups or salts thereof (27). The efficacy of the TiO_2 as a hiding pigment is reduced when TiO_2 particles are allowed to come too close together on film formation and drying. It is known that the spacing of TiO_2 and its concomitant hiding efficiency can be improved with the aid of emulsion polymers particles adsorbed on the TiO_2 particle surface (28).

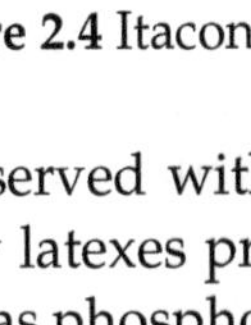

Figure 2.4 Itaconic acid.

One of the problems observed with conventional adsorbing latex technology, particularly latexes prepared using highly reactive functional monomers such as phosphoethylmethacrylate, is the formation of grit arising from the uncontrolled reaction of the reactive adsorbing latex with TiO_2. To control grit, the formulator must carefully mix the adsorbing latex with the pigment under controlled conditions to avoid flocculation, which often requires expensive high shear mixing (27).

It has been found that paints prepared using a combination of an itaconic acid-functionalized binder, a sulfonic acid-functionalized polymer, and TiO_2 show an improved hiding over itaconic acid-functionalized binders that do not include the water-soluble polymer or methacrylic acid-functionalized binders (27).

2.7.2 *Xylan-Coated Paper Laminates*

Functionalized xylans, namely a carboxymethyl xylan and a 2-dodecenyl succinic anhydride-modified xylan, were synthesized from beechwood xylan and characterized with regard to their structural properties, thermal behavior and molar mass (29).

These materials were used in the production of paper/xylan laminates for food packaging.

Films prepared from the functionalized xylans by solvent casting were applied onto paper, employing calendering molding under preselected conditions in order to produce the laminates. The obtained laminates, as well as the starting base paper and the xylan-derived films, were characterized for their mechanical and barrier properties.

These novel types of laminates demonstrated a synergistic effect with respect to their individual constituents, as the Young's modulus, tensile, tear and burst strengths of laminates were significantly improved in comparison to those of the starting paper or the xylan films (29).

The xylan-coated paper laminates showed good moisture barrier properties, reducing up to 30-fold the water vapor permeability of the paper. At the same time, despite the fairly good oxygen barrier properties of the biobased films from functionalized xylan, their application in paper laminates did not permit maintaining the same order of magnitude of oxygen transfer rates, which were, however, comparable to those reported for packaging papers coated by poly(ethylene) (PE) films (29).

2.7.3 *Surface Functionalization of Magnetic Nanoparticles*

Iron oxide nanoparticles (IONPs) have attracted significant attention for a wide range of biomedical applications (30). For the successful use of oxide nanoparticles in nanobiotechnology, surface coating and specific functionalization is critical. Many types of materials can be used in the surface coating of oxide nanoparticles for nano applications and biological applications, including organic compounds and inorganic materials.

The recent developments and various strategies for surface coating of iron oxide nanoparticles have been reviewed (30). Also, different materials used for the functionalization of iron oxide nanoparticles were discussed in detail. The design of iron oxide nanoparticles with multifunctional coatings for bioapplications is an area of considerable interest. Surface functionalization of iron oxide nanoparticles allows them to attach to various biomolecules, making them a promising candidate for biological applications.

The stability of iron oxide nanoparticles is important for their storage. Iron oxide nanoparticles are reactive toward oxidizing agents and moisture. Thus, bare iron oxide nanoparticles are usually unstable and tend to agglomerate. Therefore, surface coating and functionalization are necessary to impart colloidal stability and prevent agglomeration. To prevent aggregation and enhancing the colloidal stability, the functionalization of iron oxide nanoparticles gives rise to higher water compatibility and better advances strategies for the surface functionalization.

Magnetic nanoparticles were synthesized through the *in-situ* coprecipitation of ferrous Fe^{2+} and ferric Fe^{3+} ions from aqueous solutions in the presence of functional block copolymers (31).

In contrast to most small molecules, organic polymers attach to nanoparticles via multiple functional groups, resulting in a stronger steric repulsive force. Owing to the excellent colloidal stability of IONPs, polymer functionalized IONPs are receiving greater attention.

Polymer coating typically requires the use of active terminal groups. Reactive monomers that have been used to promote the attachment of polymer coatings to the surface of MNPs include alkoxysilanes, citric acid, c.f. Figure 2.5, bisphosphonates, and DMSA (32–34).

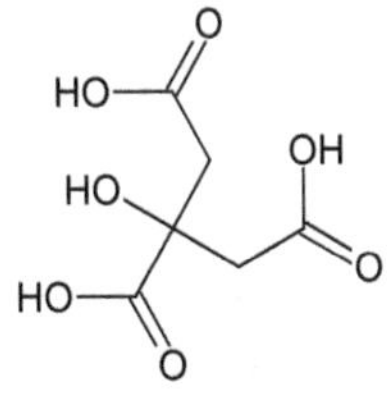

Figure 2.5 Citric acid.

The encapsulation of IONPs in a biocompatible polymer or inorganic compounds is another stabilization and modification strategy. Also, biocompatible hydrophilic shell encapsulation can be used for IONP modification. Amphiphilic ligands, water-soluble polymer matrixes, and hydrophilic inorganic materials are typical shell materials. Here, a large number of natural and synthetic biodegradable polymers, e.g., poly(aspartate) (35), polysaccharides (36), alginate, poly(ethylene glycol) (PEG) (37), chitosan (38), and copolymers such as poly(maleic anhydride-*alt*-1-octadecene) and others, can be used (30).

2.7.4 *Surface Functionalization by Thiol-ene Chemistry*

Thiol-ene-based polymer particles are traditionally prepared via emulsion polymerization in water using surfactants, stabilizers, and cosolvents. A green and simple alternative has been presented that has excellent control over particle size, while avoiding the addition of stabilizers (39).

Glycerol was used as the dispersing medium for the preparation of off-stoichiometric thiol-ene microparticles, where sizes in the range of 40 μm – 400 μm were obtained solely by changing the mixing speed of the emulsions prior to crosslinking. Control over the surface chemistry was achieved by surface functionalization of excess thiol groups via photochemical thiol-ene chemistry, resulting in a functional monolayer.

In addition, surface chain transfer free-radical polymerization was used to introduce a thicker polymer layer on the particle surface. The application potential of the system was demonstrated by using functional particles as adsorbent for metal ions and as a support for immobilized enzymes (39).

2.8 Compatibilization of Polar Polymers

The compatibilization of polar polymers, such as poly(amide) (PA) and/or ethylene vinyl alcohol with nonpolar polymers such as PE and/or poly(propylene), is commonly achieved by maleic anhydride grafted polymers. Polymer compositions containing functionalized polymers have been described (40–43).

Traditional maleic anhydride functionalized olefin-based polymers, containing high maleic anhydride content, have been used as compatibilizers between polyolefin and ethylene vinyl alcohol and/or PA in multicomponent compositions, where the maleic anhydride functionalized polyolefin compatibilizers are olefin-based polymers and ethylene vinyl alcohol and/or PA. However, in multicomponent compositions, where the interfacial surface areas are higher, the traditional maleic anhydride functionalized polyolefins can react and form a crosslinked interphase with the polar polymer (43). Articles such as films or sheets produced from such compositions typically have poor optical and mechanical properties.

Therefore, there is a need for new polymer compositions that will effectively compatibilize polyolefin/PA, polyolefin/ethylene vinyl alcohol, or, polyolefin/PA/ethylene vinyl alcohol systems, without resulting in crosslinked interphases, and which can be used to form films or sheets with improved optical and mechanical properties (43).

Functionalized derivatives of segmented or multiblock interpolymers have been described. The term *interpolymer* means a polymer prepared by the polymerization of at least two different types of monomers, i.e., both copolymer and terpolymer. The functionalized interpolymers often exhibit lower viscosities for better melt flows and lower operating temperatures in various processing applications.

These functionalized multiblock interpolymers and polymeric blends may be employed in the preparation of solid articles, such as moldings, films, sheets, and foamed objects. These articles may be prepared by molding, extruding, or other processes. The functionalized interpolymers are useful in adhesives, tie layers, laminates, and polymeric blends.

A method of polymerization consists of contacting ethylene and other monomers under addition polymerization conditions with a catalyst composition such as a (43):

1. First olefin polymerization catalyst having a high comonomer incorporation index, a
2. Second olefin polymerization catalyst having a comonomer incorporation index of most preferably less than 5% of the comonomer incorporation index of catalyst 1, and a
3. Chain shuttling agent.

Examples of catalysts are shown in Table 2.1. Examples are shown in Figures 2.6 and 2.7. Shuttling agents are shown in Table 2.2.

Table 2.1 Catalysts (43).

Compound
(*N*-(2,6-Di(1-methylethyl)phenyl amido)(2-isopropylphenyl)(α-naphthalen-2-diyl(6-pyridin-2-diyl)methane))hafnium dimethyl
(*N*-(2,6-Di(1-methylethyl)phenyl amido)(2-methylphenyl) (1,2-phenylene-(6-pyridin-2-diyl)methane))hafnium dimethyl Bis[*N*,*N*‴-(2,4,6-tri(methylphenyl)amido)ethylene diamine]hafnium dibenzyl
Bis((2-oxoyl-3-(dibenzo-1H-pyrrole-1-yl)-5-(methyl)phenyl)-2-phenoxymethyl) cyclohexane-1,2-diyl zirconium (IV) dibenzyl
1,2-Bis-(3,5-di-*tert*-butylphenylene)(1-(*N*-(1-methylethyl) imino) methyl)(2-oxoyl)zirconium dibenzyl
1,2-Bis-(3,5-di-*tert*-butylphenylene)(1-(*N*-(2-methylcyclohexyl)-imino) methyl)(2-oxoyl)zirconium dibenzyl
(*tert*-Butylamido)dimethyl(3-*N*-pyrrolyl-1,2,3,3a,7a-ε-inden-1-yl)silanetitanium dimethyl
tert-Butylamido)di(4-methylphenyl)(2-methyl-1,2,3,3a,7a-ε-inden-1-yl)silanetitanium dimethyl
(*tert*-Butylamido)di(4-methylphenyl)(2-methyl-1,2,3,3a,8a-ε-s-indacen-1-yl)silanetitanium dimethyl
Bis(dimethyldisiloxane)(indene-1-yl)zirconium dichloride

2.9 Supramolecular Polymers

Supramolecular polymers can be random or entangled coils with the mechanical properties of plastics and elastomers, but with a great capacity for processability, recycling, and self-healing, due to their reversible monomer-to-polymer transitions (44).

On the other hand, supramolecular polymers can be formed by a self-assembly among designed subunits to yield shape-persistent and highly ordered filaments. The use of strong and directional interactions among molecular subunits can achieve not only rich dynamic behavior but also high degrees of internal order, which are not known in ordinary polymers.

They can resemble, for example, the ordered and dynamic one-dimensional supramolecular assemblies of the cell cytoskeleton and possess useful biological and electronic functions (44).

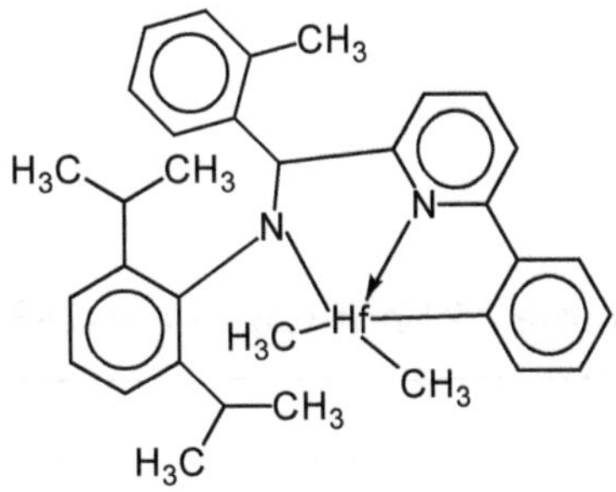

Figure 2.6 Bis[*N,N'''*-(2,4,6-tri(methylphenyl)amido)ethylene diamine]-hafnium dibenzyl.

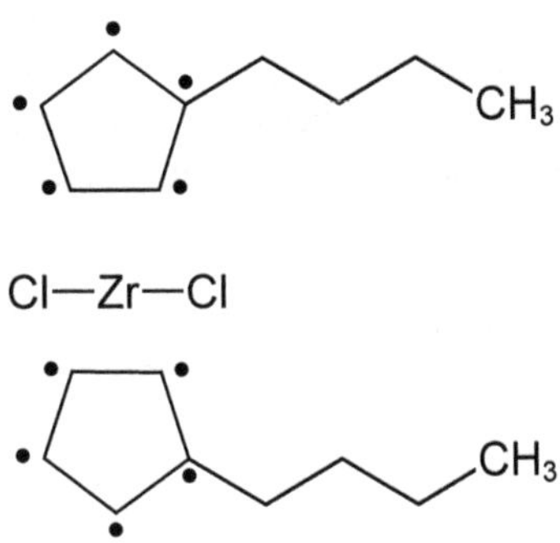

Figure 2.7 Bis(dimethyldisiloxane)(indene-1-yl)zirconium dichloride.

Table 2.2 Shuttling agents (43).

Compound
Diethylzinc
Di(*i*-butyl)zinc
Di(*n*-hexyl)zinc
Triethylaluminum
Trioctylaluminum
Triethylgallium
i-Butylaluminum bis(dimethyl(*tert*-butyl)siloxane)
i-Butylaluminum bis(di(trimethylsilyl)amide)
n-Octylaluminum di(pyridine-2-methoxide)
Bis(*n*-octadecyl)*i*-butylaluminum
i-Butylaluminum bis(di(*n*-pentyl)amide)
n-Octylaluminum bis(2,6-di-*tert*-butylphenoxide)
n-Octylaluminum di(ethyl(1-naphthyl)amide)
Ethylaluminum bis(*tert*-butyldimethylsiloxide)
Ethylaluminum di(bis(trimethylsilyl)amide)
Ethylaluminum bis(2,3,6,7-dibenzo-1-azacycloheptaneamide)
n-Octylaluminum bis(2,3,6,7-dibenzo-1-azacycloheptaneamide)
n-Octylaluminum bis(dimethyl(*tert*-butyl)siloxide
Ethylzinc(2,6-diphenylphenoxide)
Ethylzinc(*tert*-butoxide)

2.9.1 *Supramolecular Nanofiber Networks*

Self-assembling supramolecular nanofibers are of technical importance to both nanotechnology and materials science (45).

The low-complexity sequence domain of a fused sarcoma protein was investigated (45). This is an essential cellular nuclear protein with slow kinetics of amyloid fiber assembly. It can construct random copolymer-like, multiblock, and self-sorted supramolecular fibrous networks with distinct structural features and fluorescent functionalities.

The usage of these networks was demonstrated in a templated, spatially controlled assembly of ligand-decorated gold nanoparticles, quantum dots, nanorods, deoxyribonucleic acid (DNA) origami, and hybrid structures. Owing to the distinguishable nanoarchitectures of these nanofibers, this assembly is structure-dependent.

By coupling a modular genetic strategy with a kinetically controlled complex supramolecular self-assembly, it could be demonstrated that a single type of protein molecule can be used to engineer diverse one-dimensional supramolecular nanostructures with distinct functionalities (45).

2.9.2 *Supramolecular Biomaterials*

The translation of supramolecular polymers into multicomponent functional biomaterials for regenerative medicine applications have been collected in a review (46).

Several examples of functional supramolecular biomaterials reaching the clinic have been reported. The basic concept of many of these supramolecular biomaterials is based on their ability to adapt to cell behavior as a result of dynamic noncovalent interactions.

A variety of material preparation methods, i.e., 3D printing, melt spinning and electrospinning, can be realized, which hold great promise for the production of functional supramolecular biomaterials that can be applied in the field of regenerative medicine (47).

The modular character of supramolecular elastomeric materials allows the introduction of complex (bio)functionality. As a result of the interesting properties of the thermoplastic elastomers, they have been suggested to be applied in a broad range of regenerative medicine applications, i.e., in the cardiovascular field.

Several functional amino acids as well as adhesive peptides have been incorporated into peptide amphiphiles to introduce biofunctionality, and were shown to form 1D fibers despite the chemical modifications (48).

2.10 Microgels

Microgels are macromolecular networks swollen by the solvent in which they are dissolved (49). They are unique systems that are distinctly different from common colloids such as rigid nanoparticles, flexible macromolecules, micelles, or vesicles. The size of microgel networks is in the range of several micrometers down to nanometers.

Microgels have the possibility to introduce chemical functionalities at different positions. The combination of architectural diversity and compartmentalization of reactive groups enables a short-range coexistence of otherwise unstable combinations of chemical reactivity. The open microgel structure is beneficial for uptake-release purposes of active substances.

In addition, the openness allows site-selective integration of active functionalities like reactive groups, charges, or markers by post-modification processes. The unique ability of microgels to retain their colloidal stability and swelling degree both in water and in many organic solvents allows the use of different chemistries for the modification of the structure of the microgel.

Thus, microgels allow combining features of chemical functionality, structural integrity, macromolecular architecture, adaptivity, permeability, and deformability in an unique way (49).

2.11 Polymers from Sulfur

Polymers are among the most important mass-produced materials on the planet, however they are largely derived from a finite supply of petrochemicals. To ensure the sustainable production of polymers and functional materials, alternative feedstocks are required (50).

Sulfur is a by-product of the petroleum industry. The conversion of sulfur into useful polymers and related materials is an advance in waste valorization. In addition, the copolymerization of sulfur with renewable monomers represents an additional contribution to sustainability. These reactions are often solvent-free and benefit from full atom economy.

In order to produce polymers directly from sulfur, several challenges appear. The main limitation has been the instability of polysulfides made by the ring-opening polymerization of sulfur.

When elemental sulfur is heated above its floor temperature of 159°C, the S–S bond homolysis provides thiyl radicals that attack and open the ring of another molecule. The polymerization is then propagated by repeated ring-opening and S–S bond formation between S_8 and the growing polysulfide chain.

However, the reaction is reversible and the terminal thiyl radicals of the polysulfide can depolymerize and expel sulfur compounds. To provide stable polysulfide polymers, the thiyl radicals must be quenched before depolymerization.

The addition of the linear polymeric sulfur to polyenes results in a branching of the polymeric chains. In the termination events, the thiyl radicals are thought to be quenched by at least two different mechanisms. In one case, intramolecular recombination of thiyl radicals would provide stable polysulfide loops.

When hydrogen atoms are available, e.g., allylic and benzylic hydrogen atoms from the alkene C-monomer, hydrogen atom abstraction may convert the thiyl radical into a thiol.

Besides alkene and alkyne crosslinking agents, other functional groups can react with sulfur, in which alternative mechanisms are operative in the inverse vulcanization. It has been illustrated how polythiols such as trithiocyanuric acid, c.f. Figure 2.8, can be used to prepare sulfur-rich polymers by reaction with elemental sulfur (51). Also, aromatic thiol crosslinking agents were assessed for inverse vulcanization (52). These procedures are of interest for lithium-sulfur batteries.

Several crosslinking agents for inverse vulcanization are summarized in Table 2.3. Some crosslinking agents are shown in Figure 2.9.

Also, a method was presented to form polysulfide foams from the polymers using supercritical carbon dioxide (68).

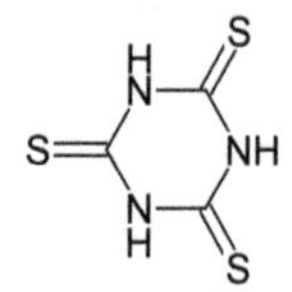

Figure 2.8 Trithiocyanuric acid.

Table 2.3 Crosslinking agents for inverse vulcanization (50).

Compound	Reference
1,3,5-Triisopropenylbenzene	(53)
Styrene	(54)
Methylstyrene	(55)
Divinylbenzene	(56–58)
Bismaleimide	(57)
Oleylamine	(59)
Allyl-terminated poly(3-hexylthiophene-2,5-diyl)	(60)
Styrenic functional 3,4-propylenedioxythiophene	(61)
1,3-Diethynylbenzene	(62)
1,4-Diphenylbutadiyne	(63)
D-Limonene	(64)
Natural dienes	(65)
Poly(isoprene)	(66)
Composites from Vegetable Oils	(67)

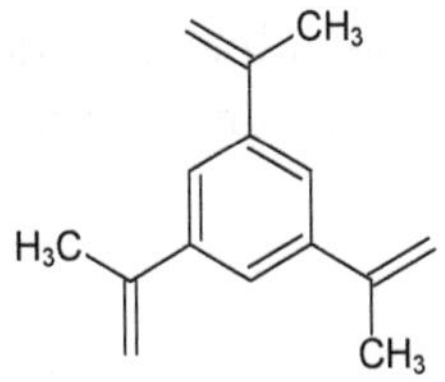

1,3,5-Triisopropenylbenzene

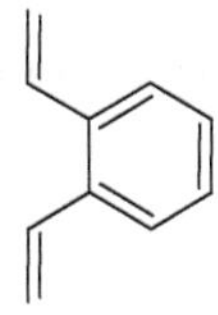

o-Divinylbenzene

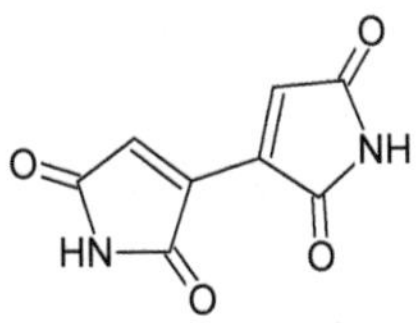

Bismaleimide

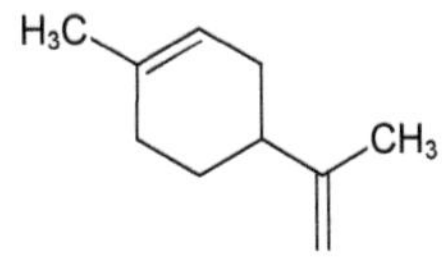

D-Limonene

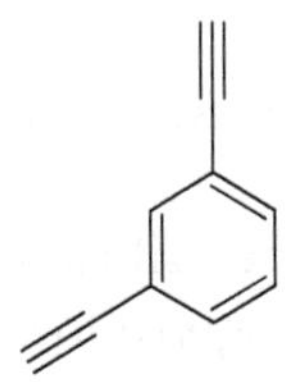

1,3-Diethynylbenzene

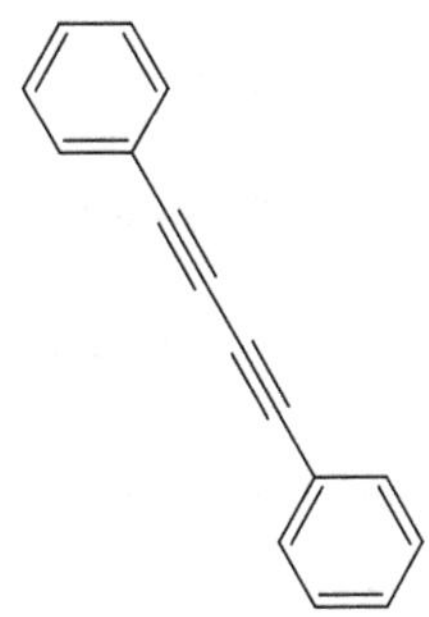

1,4-Diphenylbutadiyne

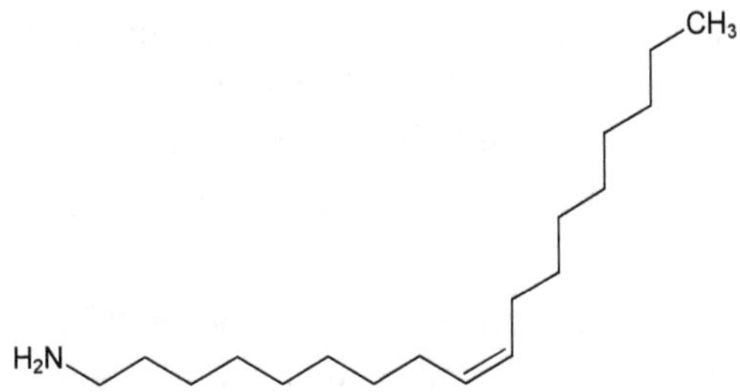

Oleylamine

Figure 2.9 Crosslinking agents.

These foams from polysulfides are fabricated by the reaction of sulfur and dicyclopentadiene and renewable terpenes such as myrcene, farnesol and farnesene (69). These compounds are shown in Figure 2.10.

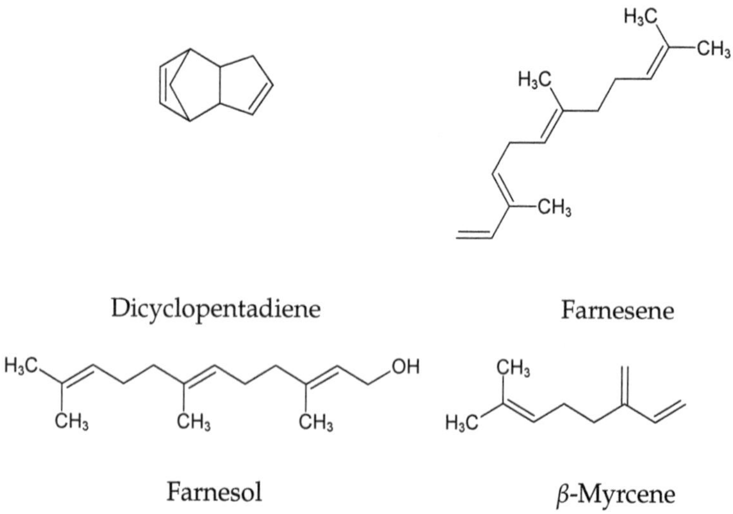

Figure 2.10 Foam reagents.

The high surface area imparted by supercritical carbon dioxide foaming or using sodium chloride as a porogen allowed the efficient sequestration of inorganic mercury from water.

Polysulfides that were prepared by inverse vulcanization have been explored in diverse areas of sustainability, including power generation, power storage, photocatalytic water splitting, and environmental remediation (50). Hole-transport materials in solid-state dye-sensitized solar cells have been prepared (70).

Polysulfides are redox active and useful for cathode materials for lithium-sulfur batteries (71,72)

Nanowires have been prepared from polysulfides and their use for the photochemical splitting of water with visible light was explored to generate hydrogen, a clean-burning fuel (73). Poly(sulfur-*random*-1,3-diisopropenylbenzene) copolymer nanowires were fabricated using anodic aluminum oxide membranes as templates. The structure of the copolymer is shown in Figure 2.11.

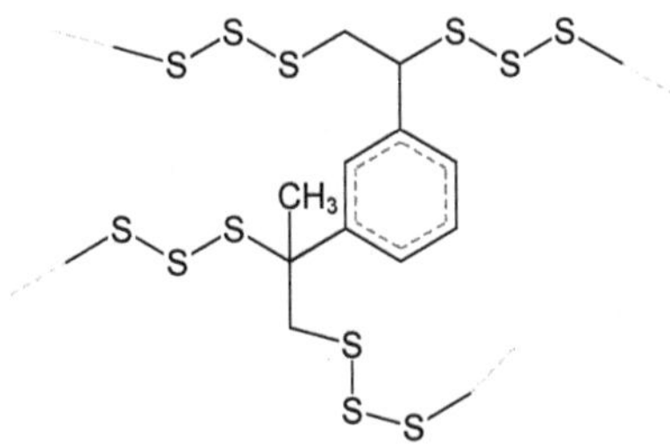

Figure 2.11 Poly(sulfur-random-1,3-diisopropenylbenzene) (73).

The template was removed by etching with sodium hydroxide, providing the polysulfide nanowires. The photocatalytic activity of these polysulfide nanowires was superior to bulk sulfur, a feature attributed in part to their high surface area. This study is an important report of the ways in which inexpensive sulfur can be converted into a valuable catalyst that can harness visible light for the generation of clean fuels (50).

2.12 Aliphatic Polyesters

2.12.1 Aliphatic Polyesters with Pendant Carboxyl Groups

Aliphatic polyesters with pendant carboxyl groups can be prepared by the ring-opening polymerization of cyclic esters with benzyl-protected carboxyl groups.

For example, a poly(α-malic acid) was prepared as a carboxyl functional analogy of poly(lactic acid) (PLA) by the ring-opening polymerization of malide dibenzyl ester, c.f. Figure 2.12, followed by acid deprotection (74).

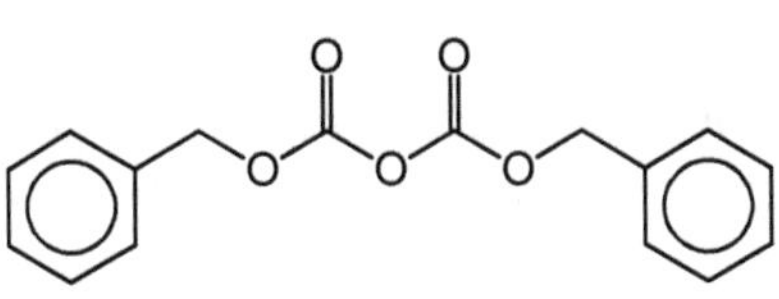

Figure 2.12 Malide dibenzyl ester.

Also, the synthesis of poly[(α-malic acid)-*alt*-(glycolic acid)], a glycolide-based poly(ester) with pendant carboxylic acid, was reported (75). Here, the ring-opening polymerization of 3(S)-[(benzyloxycarbonyl)-methyl]-1,4-dioxane-2,5-dione, c.f. Figure 2.13, is followed by debenzylation. These aliphatic copolyesters are hydrolyzed more rapidly than PLA.

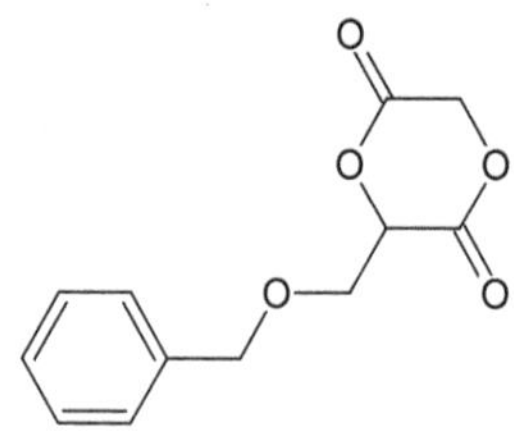

Figure 2.13 3-[(Benzyloxycarbonyl)-methyl]-1,4-dioxane-2,5-dione.

Side-chain-functionalized lactide analogues could be prepared from commercially available amino acids (76). The thus functionalized cyclic monomers could be homopolymerized and copolymerized with lactides and then quantitatively deprotected, forming functional PLA-based materials with amino, hydroxyl or carboxyl side chains.

Sequential and simultaneous cationic copolymerization reactions of lactide with commercial functional epoxides have been studied (77). The functional epoxides were allyl glycidyl ether, glycidyl propargyl ether, and epichlorohydrin. These compounds are shown in Figure 2.14.

The reaction was catalyzed by a protic acid in a one-pot process. Block and gradient poly(lactide)-based medium-molecular-weight copolymers with several alkene, alkyne, or chloromethyl groups at one chain end or with gradient distribution along the polymer chain could be successfully synthesized.

Furthermore, functional poly(lactide)s were subjected to thiol-ene addition, cycloaddition of azide, or reaction with tertiary amine, respectively, to demonstrate the reactivity of the introduced functional groups (77).

The synthesis and the polymerization of benzyl malolactonate has been described (78).

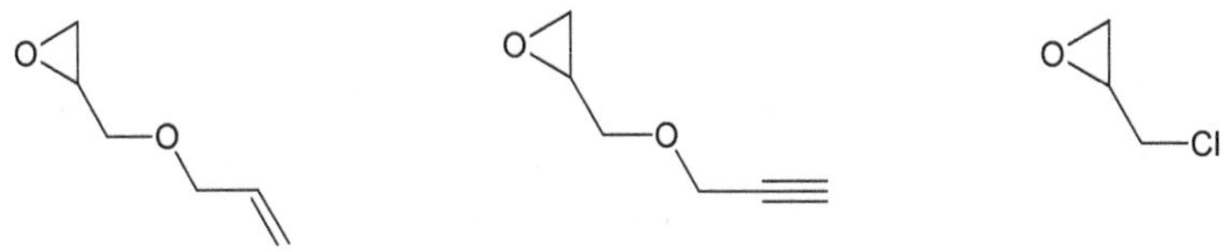

Allyl glycidyl ether Glycidyl propargyl ether Epichlorohydrin

Figure 2.14 Functional epoxides.

The synthesis starts from optically active aspartic acid. Chiral benzyl malolactonate is a β-substituted β-lactone monomer, which can be anionically polymerized using triethylamine as the initiator to yield poly(benzyl-β-malate), which is an optically active, semi-crystalline polymer. The benzyl protecting groups could be readily removed by catalytic hydrogenolysis. The cleavage of the protecting benzyl ester groups yields optically active poly(β-malic acid) (78).

A poly(*l*-lactide-co-β-malic acid) with a high molecular weight could be synthesized by the copolymerization of *l*-lactide and β-benzyl malolactonate (79). β-Benzyl malolactonate was synthesized from bromosuccinic acid and benzyl alcohol in the presence of trifluoroacetic acid anhydride. β-Benzyl malolactonate is shown in Figure 2.15.

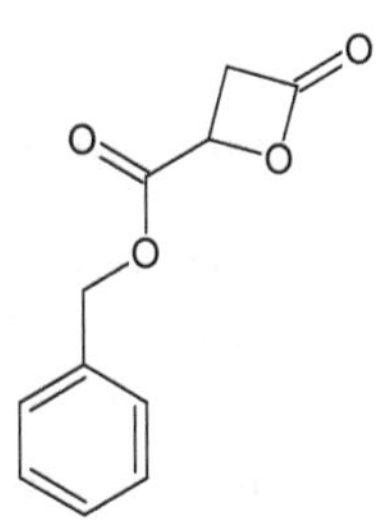

Figure 2.15 β-Benzyl malolactonate.

The degradation products of this polymer are the nontoxic products lactic acid and malic acid, which are the intermediates in the saccharide metabolism in the human body. The hydrophilicity of the deprotected product increases with increasing malic acid content (79).

Poly(caprolactone) (PCL) with pendant carboxylic acid groups were prepared by the ring-opening polymerization of benzyl γ-(ϵ-caprolactone) carboxylate or *tert*-butyl-γ-(ϵ-caprolactone) carboxylate followed by acid deprotection (80)

2.12.2 *Aliphatic Polyesters with Amino Groups*

Amino functionalized PCL was synthesized by the ring-opening polymerization of 4-trifluoroacetyl-7-oxo-1,4-oxazaperhydroepine, c.f. Figure 2.16, followed by the deprotection with $NaBH_4$ (80).

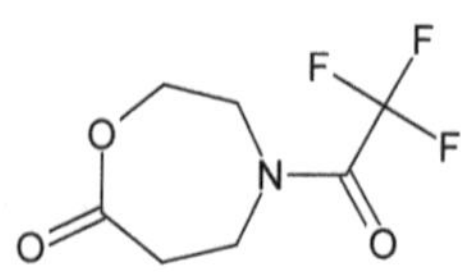

Figure 2.16 4-Trifluoroacetyl-7-oxo-1,4-oxazaperhydroepine.

Also, the preparation of an aliphatic polyester was reported that bears lateral amino groups by the anionic ring-opening polymerization of *N*-tritylated serine β-lactones (81). Polymers with molecular weights up to 30 *k Dalton* have been obtained. The removal of the protective group results in a polyserine ester.

2.12.3 *Aliphatic Polyesters with Chloride Groups*

A chloro-substituted four-membered lactone, α-chloromethyl-α-methyl-β-propionolactone, was synthesized (82).

This compound was polymerized and copolymerized with various amounts of ε-caprolactone. Aluminium triisopropoxide was used as initiator (82). These compounds are shown in Figure 2.17.

The pendant chloromethyl groups of the copolymers were converted into quaternary ammonium salts by the reaction with pyridine, which increased the hydrophilicity of the copolymer.

In another study, both electrophiles like trifluoroacetic acid and nucleophiles like triethylamine and pyridine, as well as organometallic compounds such as stannous octoate, aluminium triiso-

α-Chloromethyl-α-methyl-β-propionolactone ε-Caprolactone

Figure 2.17 Components for a four-membered lactone.

propoxide and tetrabutyl orthotitanate, were found to be effective initiators (83). The crystallinity of the polymer was found to vary with the types of initiator used.

2.12.4 *Aliphatic Polyesters with Keto or Hydroxyl Groups*

Aliphatic polyesters bearing keto groups were synthesized by the ring-opening polymerization of 5-ethylene ketal ε-caprolactone followed by deprotection (84–87).

The keto groups of the copolymers could be efficiently reduced into hydroxyl groups by using $NaBH_4$ in a mixture of dichloromethane and ethanol at room temperature without any apparent chain degradation. This procedure resulted in aliphatic polyesters with pendant hydroxyl groups (85).

PCL containing pendant hydroxyl groups were prepared by the ring-opening polymerization of ε-caprolactone monomer with triethylsilyloxy pendant groups that could be selectively deprotected into hydroxyl groups under mild conditions (88).

Furthermore, the synthesis and polymerization of γ-benzyloxy-ε-caprolactone and γ-2,2-bis(phenyldioxymethyl)propionate-ε-caprolactone was reported (80).

The catalytic hydrogenolysis of the benzyl protecting group of the products afforded PCL with pendant hydroxyl or bishydroxyl groups, respectively.

2.12.5 *Aliphatic Polyesters with Bromide Groups*

The preparation of aliphatic polyesters with pendant bromide groups by the ring-opening polymerization of a bromo-substituted cyclic ester, γ-(2-bromo-2-methyl propionyl)-ε-caprolactone (BMPCL) with a pendant activated alkyl bromide functional group has been described (89). The pendant activated alkyl bromide group could initiate a controlled ATRP of methyl methacrylate.

Therefore, PCL-graft-poly(methyl methacrylate) copolymers could be obtained in a simple one-step approach by the concurrent polymerization of an ε-caprolactone, BMPCL, and methyl methacrylate with an appropriate initiator for the ring-opening polymerization and a catalyst for the ATRP.

2.12.6 *Aliphatic Polyester Biopolymers*

The copolymerization of morpholine-2,5-dione derivatives with lactide or lactones is a convenient way to prepare aliphatic biopolymers bearing reactive groups (83). Morpholine-2,5-dione is shown in Figure 2.18.

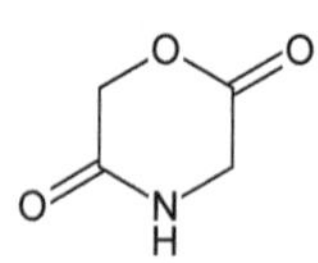

Figure 2.18 Morpholine-2,5-dione.

It was demonstrated that the ring-opening polymerization of either ε-caprolactone or *d,l*-lactide with morpholine-2,5-dione derivatives could protect functional substituents, such as benzyl-protected carboxylic acid, benzyloxycarbonyl-protected amine and *p*-methoxy-protected thiol groups. Polyesteramides with pendant carboxylic acid groups, pendant amine groups, or pendant thiol groups were obtained after the deprotection of the copolymers (90).

2.12.7 *Lactide-Functionalized Polymers*

The depletion of fossil fuels from which the majority of polymers are derived, combined with supply chain instability and cost fluctuations of feed chemicals used to make these polymers, are driving the development and utilization of biobased plastics for commodity applications (91). PLA, derived from starch and sugars, is a particularly appealing biobased plastic that is inexpensive and already being produced in large commercial quantities.

PLA is capable of replacing many petroleum-derived polymers in some applications. However, several of its properties, such as low heat distortion temperature, high water adsorption, low flame retardancy, and low toughness, exclude the use of PLA in many applications. Much effort has been dedicated to directly incorporate chemical functionalities into the backbone of PLA in order to tailor its properties.

Lactide-functionalized polymers were prepared from a brominated lactide monomer using ATRP (91). Lactide is a commercially available biobased cyclic ester monomer that can be obtained from biomass. Lactide is the cyclic diester of lactic acid. A lactide may be prepared by heating lactic acid in the presence of an acid catalyst.

Many monomers, such as styrenic, vinylic, acrylic and others, can be used in radical polymerization. As a brominated lactide initiator, 3-bromo-3,6-dimethyl-1,4-dioxane-2,5-dione can be used. 3-Bromo-3,6-dimethyl-1,4-dioxane-2,5-dione can be prepared by reacting a lactide with *N*-bromosuccinimide in the presence of dibenzoyl peroxide. Then, a lactide-functionalized polymer can be obtained by ATRP of a vinylic monomer. The synthesis route is shown in Figure 2.19.

As suitable catalytic complex, a CuBr *N,N,N′,N″,N″*-pentamethyldiethylenetriamine composition may be used. Other organic ligands include 4,4′-dinonyl-2,2′-bipyridine and 1,1,4,7,10,10-hexamethyltriethylenetetramine. These compounds are shown in Figure 2.20.

Such materials may be used as macromonomers. In another application, the lactide end group of lactide-functionalized polymers may be modified through hydrolysis or esterification to liberate other useful functional end groups, e.g., hydroxyl, carboxylic acid, alkenyl, alkynyl, and other end groups, from which subsequent chemistry can be performed. Also, the lactide-functionalized polymers may be used to functionalize surfaces, particles, or other functional polymers via a facile reaction of the lactide end group to hy-

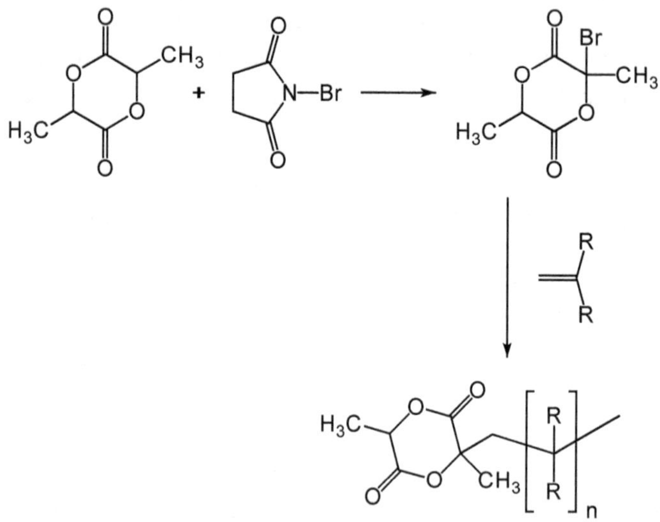

Figure 2.19 Synthesis of lactide-functionalized polymers (91).

droxyl or amine groups on the surface, particle, or other functional polymers (91).

2.13 Graphitic Carbon Nitride Polymers

Mesoporous micro-/nanostructures acting as supports for catalysts or used directly in catalysis reactions generally show interesting performances that could lead to great potential for their application (92). In the past few decades, extensive efforts have been devoted to the exploration and enrichment of graphitic carbon nitride-based research. Especially, mesoporous graphitic C_3N_4 with a controllable porosity and electronic/atomic structure can bring to bear unique physicochemical properties and has been widely applied in the fields of photocatalysis, adsorbents, sensors and chemical templates.

The significant advances in functional mesoporous graphitic C_3N_4 polymers have been reviewed (92). This includes general synthesis strategies and growth mechanisms, modifications of electronic/atomic structures and interfacial properties (such as exfoliation, doping and hybridizing), as well as their current applications.

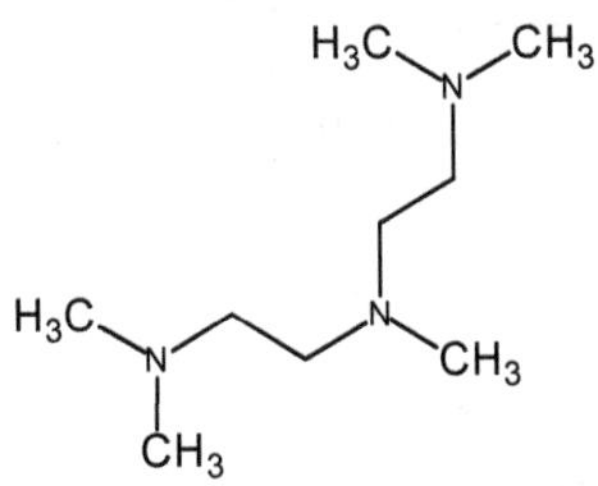

N,N,N',N",N"-Pentamethyldiethylenetriamine

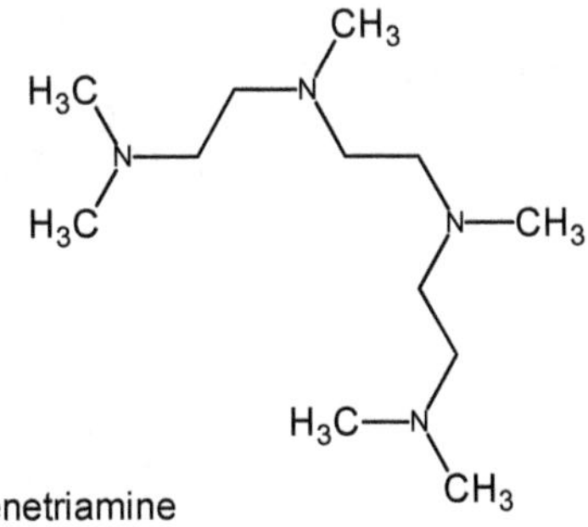

1,1,4,7,10,10-Hexamethyltriethylenetetramine

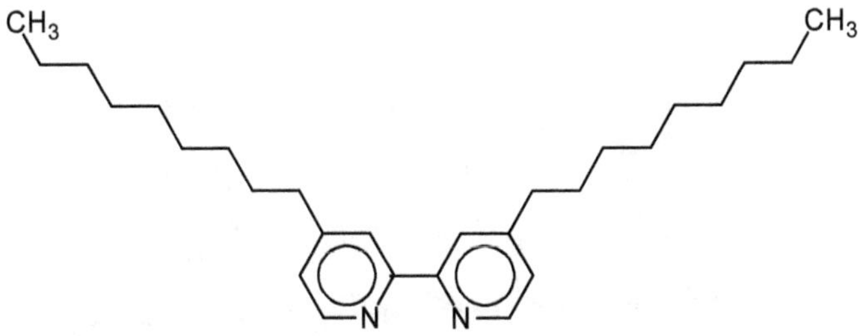

4,4′-Dinonyl-2,2'-bipyridine

Figure 2.20 Organic ligands (91).

Mesoporous graphitic C_3N_4 can be synthesized by nanocasting/replication of silica nanospheres or mesoporous silica objects, c.f. Figure 2.21. The first mesoporous graphitic C_3N_4 nanostructures were prepared through the replication of silica sols with an average size of 12 *nm* by using thermal condensation of cyanamide powders (93,94).

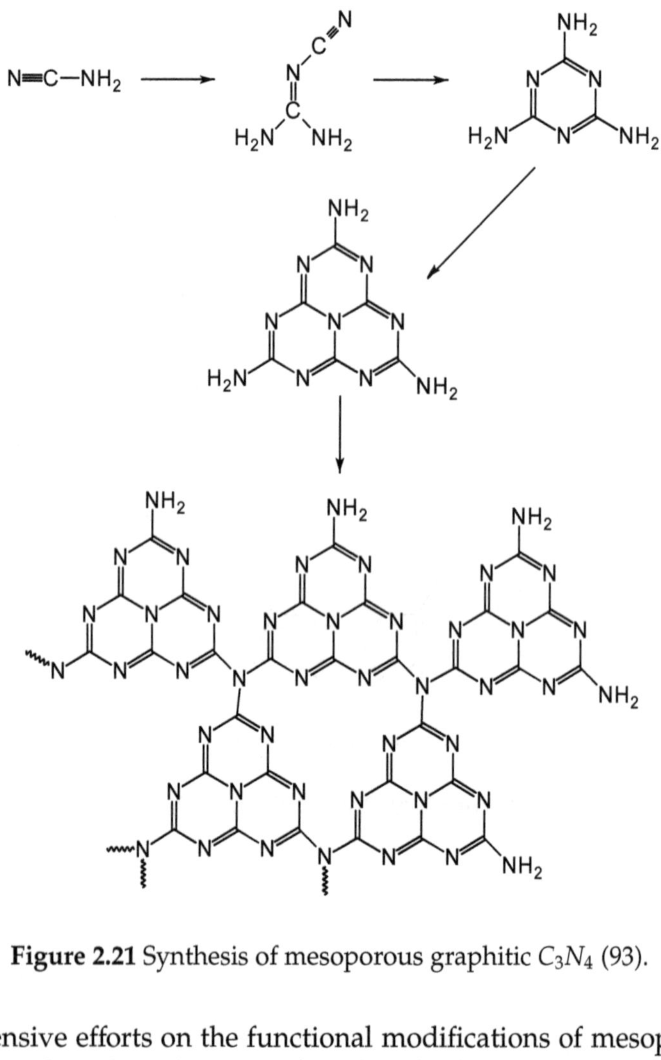

Figure 2.21 Synthesis of mesoporous graphitic C_3N_4 (93).

Extensive efforts on the functional modifications of mesoporous graphitic C_3N_4 have been employed in the fields of photocatalysis, electrocatalysis and photoelectrocatalysis (95–102).

2.14 Functionalized Buckminster[60]fullerene

Microporous polymers could be constructed through a covalent cross linkage of the buckyball and linear dialkyne through a copper-mediated, one-pot synthesis (103). So, a novel functionalization-polymerization pathway was described whereby a copper-mediated free radical species, generated *in situ* triggered multiple additions of the ditopic alkyne to C_{60}. The multifunctionalized C_{60} species act as nodes that afforded microporous networks on extension through proper bridges.

Solid-based spectroscopy techniques allowed the identification of the chemical composition of the polymers, while gas sorption measurements were utilized to probe the microporosity, surface area, and pore size distribution of the constructed solids. Electron paramagnetic resonance spectroscopy demonstrated the presence of free radicals at the polymerization reaction.

In addition to a relatively unstable diisopropyl nitroxide radical, electron paramagnetic resonance revealed a quasi-stable, C_{60}-derived, organic radical that exhibits an unusually intense microwave absorption and long lifetime at room temperature of more than 5 *d* (103).

2.15 Functional Semi-fluorinated Polymers

The construction of an organic photocatalysis system has been reported (104). The system contains a photoredox catalyst/additive for an organocatalyzed step transfer-addition and radical-termination polymerization irradiated by blue LED light at room temperature.

As catalyst, 2,4,6-tri-(*p*-methoxyphenyl) pyrylium tetrafluoroborate and Eosin Y was mainly used for the construction of the organic photoredox catalysis system. These compounds are shown in Figure 2.22. The organocatalytic polymerization strategy provides a simple and efficient approach to functional semi-fluorinated polymers because of its mild reactions conditions, high tolerance to functional groups and metal-free residues in the polymer product.

Different types of α,ω-diiodoperfluoroalkanes and α,ω-unconjugated dienes were efficiently copolymerized through organocatalyzed step transfer-addition and radical-termination polymeriza-

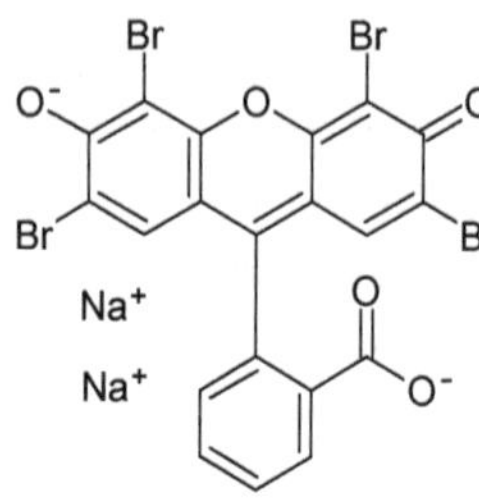

Eosin Y

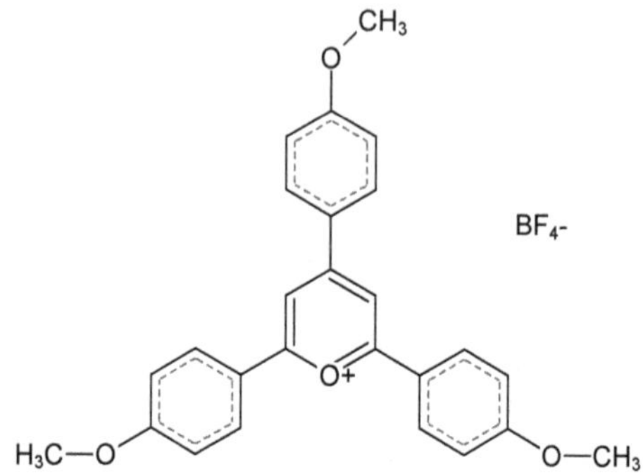

2,4,6-Tri-(*p*-methoxyphenyl) pyrylium tetrafluoroborate

Figure 2.22 Organic photoredox catalysts.

tion. Various kinds of functional semi-fluorinated polymers were generated, including polyolefins and polyesters. The process is affected by several factors: Solvents, additives, and the feed ratio of the monomers.

After optimization of all these components, the polymerization efficiency could be greatly improved, generating polymers with both relatively high yield and molecular weight. Considering the mild reaction condition, easy operation process, and free-of-metal-catalyst residues in the polymer product, the organocatalytic polymerization strategy provides a simple and efficient approach to functional semi-fluorinated polymer materials and may open up their application in high-tech fields (104).

2.16 Hydroxyl-Terminated Poly(butadiene)

Poly(butadiene) with the simultaneous presence of hydroxyl end groups and pendant carboxyl side groups could be synthesized by the combination of living anionic polymerization and a blue light photocatalytic thiol-ene click reaction (105).

Initially, a hydroxyl-terminated poly(butadiene) with a high content of 1,4-butadiene units was synthesized by living anionic polymerization using a hydroxy-protected initiator, followed by deprotection. Then, the double bonds in the backbone of hydroxyl-terminated poly(butadiene) were modified into pendant carboxyl side groups by a blue light photocatalytic thiol-ene click reaction using $Ru(bpy)_3Cl_2$ and *p*-toluidine as photoredox catalysts. The blue light photocatalytic thiol-ene reaction exhibited a high efficiency, and all the unsaturated bonds were modified after reacting for 5 *h*; despite this in-chain double bonds in 1,4-butadiene units are much less reactive than vinyl groups in 1,2-butadiene units. When all the double bonds were modified, the structure of resulting polymer can be deemed as carboxyl-functionalized hydroxyl-terminated PE.

Furthermore, the solution properties of resulting polymer have been investigated by various technologies. The results showed that carboxyl-functionalized hydroxyl-terminated PE experiences a dissolution-assembly-aggregation transition process where the pH value of the solution decreased from 12.0 to 2.0. The mechanism of this pH-triggered phase transition behavior was supposed to depend on the ionization of carboxyl side groups at varying pH values (105).

2.17 Poly(carbonate)s

Aliphatic polyesters and copolyesters are among the most commonly used degradable materials for the preparation of clinical devices (106). In this field, aliphatic poly(carbonate)s are good materials because they possess functionalizable side chains, e.g., –OH, $-NH_2$, –COOH, and others, that can easily meet the need for functionalization of biomaterials. Moreover, aliphatic poly(carbonate)s have good biocompatibility, low toxicity, and good biodegradability (107)

The mechanisms of cationic, anionic, coordination and enzymatic polymerization of cyclic carbonates have been reviewed (108). Here, polymers with well-defined structures and oligomers with reactive pendant and end groups can be obtained. The reactions of cyclic carbonates with different nucleophilic reagents leading to products with carbon dioxide retention or with decarboxylation have been reported. Also, the synthesis and polymerization of spiroorthocarbonates, bicyclic acetals, which are the intermediates in the reaction of cyclic carbonates with cyclic ethers, has been detailed.

High molecular weight aliphatic poly(carbonate)s can be prepared by the ring-opening polymerization of cyclic carbonates (109).

The most commonly used cyclic carbonates for ring-opening polymerization are five- and six-membered cyclic monomers. Polymerization of five-membered cyclic aliphatic carbonates can produce poly(ester-carbonate)s with a content of units lower than 50 *mol%* through a partial decarboxylation regardless of the initiator and reaction conditions (108).

In contrast to five-membered cyclic carbonates, six-membered cyclic carbonates can be easily polymerized and copolymerized with various heterocyclic monomers to form poly(carbonate)s without any decarboxylation under proper conditions (106).

2.18 Poly(styrene)s

Several routes for the synthesis of functional poly(styrene) (PS) materials have been described in the literature (110).

Functional groups can be introduced directly in the reactive medium, as described by the *in-situ* functionalization pathway. This includes statistical copolymerization with other monomers as well as the catalytic chain transfer to a chain transfer agent. Statistical copolymerization implies the design of highly versatile catalytic systems, able to selectively polymerize a large variety of monomers.

Another method of functionalization consists in the derivatization of the polymer once formed. This includes sequential block copolymerization, and the chemical modification of the polymer. The former implies the design of catalytic systems that enable a living polymerization. The living syndiospecific polymerization of styrene was reported in 2001 (111).

The functionalization of syndiotactic PS with succinic anhydride was performed via a Friedel-Crafts reaction in a heterogeneous process using carbon disulfide as dispersing agent and aluminum chloride as catalyst (112). The optimum reaction conditions were found at 30°C with a molar ratio of aluminum chloride to succinic anhydride of 3. The degree of succinoylation reached 5.9 *mol%*. The melting temperature, crystallization temperature, and degree of crystallinity of the succinoylated polymers decrease with increasing the degree of succinoylation, while the glass transition temperature increases.

Acetylation of syndiotactic PS was reported via the heterogeneous system using carbon disulfide as the dispersing medium and acetyl chloride as acetylating agent (113). The acetylation reaction can be carried out to a degree of functionalization as high as 43 *mol*%. As in the case of the succinoylation functionalization, the degree of crystallinity is lower in the presence of acetyl groups, while the glass transition temperature increases.

The benzoylation of syndiotactic PS was reported using carbon disulfide as the dispersing medium, and benzoyl chloride and anhydrous aluminum chloride as benzoylating agent and catalyst, respectively (114). The degree of benzoylation goes up to 36% at 30°C. The incorporation of benzoyl groups into syndiotactic PS decreases the melting point and the degree of crystallinity, while the glass transition temperature increases.

Syndiotactic PS can be sulfonated using acyl sulfate complexes prepared from acetic anhydride, lauric acid or caproic anhydride associated with sulfuric acid in 1,2,4-trichlorobenzene (115,116). The sulfonation efficiency increases significantly when anhydrides with long aliphatic groups are used to complex the sulfuric acid. The effects are shown in Table 2.4.

Table 2.4 Complexation effects (115, 116).

Complexing agent	Reaction time/[*h*]	Sulfonation efficiency/[%]
Acetic anhydride	1	20
Lauric acid	1	30
Lauric acid	3	30
Caproic anhydride	1	45

The same reaction was also reported with acetyl sulfate in a mixture of 1,1,2-trichloroethane and chloroform (117).

The controlled activation/borylation of aromatic C–H bonds of syndiotactic PS was accomplished using an iridium-based catalyst (118). The functionalization allowed the introduction of a boronate ester group with efficiency close to 42 *mol*%. The concentration of the boryl group could be easily controlled by changing the ratio of boron reagent to monomer unit. With a boryl group ratio of 5.9%, the functional polymer was found to be amorphous. Also, the aromatic borate groups in the polymer could be further substituted by other functional groups.

2.19 Alkyne-Functional Polymers

Alkyne-functional polymers synthesized by ATRP exhibit bimodal molecular weight distributions. These indicate the occurrence of some undesirable side reactions. By modeling the molecular weight distributions obtained under various reaction conditions, it was shown that the side reaction is alkyne-alkyne coupling, i.e., Glaser coupling (119).

Glaser coupling is a type of coupling reaction. It is by far the oldest acetylenic coupling and is based on cuprous salts, like copper(I) chloride or copper(I) bromide, and an additional oxidant like oxygen (120). The reaction was first reported by Carl Andreas Glaser in 1869. An example is shown in Figure 2.23.

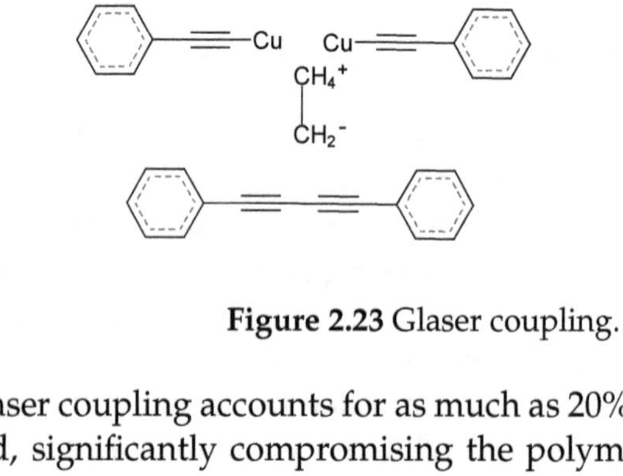

Figure 2.23 Glaser coupling.

Glaser coupling accounts for as much as 20% of the polymer produced, significantly compromising the polymer functionality and undermining the success of subsequent click reactions in which it is used. Glaser coupling does not occur during ATRP, rather during postpolymerization workup upon first exposure to air.

Two strategies have been reported that effectively eliminate these coupling reactions without the need for a protecting group for the alkyne-functional initiator (119):

1. Maintaining low temperature post-ATRP upon exposure to air followed by immediate removal of the copper catalyst, and
2. Adding excess reducing agents post-ATRP, which prevent the oxidation of the Cu(I) catalyst required by the Glaser coupling mechanism.

Post-ATRP Glaser coupling was also influenced by the ATRP synthesis ligand used. The order of ligand activity for catalyzing Glaser coupling was: linear bidentate > tridentate > tetradentate.

It was found that Glaser coupling is not problematic in activators regenerated by electron transfer-atom transfer radical polymerization (ARGET-ATRP) of alkyne-terminated polymers, because a reducing agent is present during polymerization. However, the molecular weight distribution is broadened compared to ATRP due to the presence of oxygen. Glaser coupling can also occur for alkynes held under click reaction conditions but again can be eliminated by adding appropriate reducing agents (119).

2.20 Polymers from Renewable Plant Oils

Plant oils are one of the most ideal chemical feedstocks to replace fossil resources for a variety of industrial chemicals in the polymer industry. The recent progress on plant oil-based functional polymers and composites has been reviewed (121).

An environmentally benign coating system has been demonstrated using plant oils as the starting substrate for the preparation of artificial urushi with high hardness and a high gloss surface. Epoxidized plant oils, mainly epoxidized soybean oil, are polymerized with designed inorganics, cellulose, and biodegradable aliphatic polyesters to yield functional biobased polymers and composites.

Castor oil can be used as the core of branched PLA to improve the physical properties of biobased polymers. These plant oil-based materials are significant for the molecular design of functional and high-performance biobased products for industrial applications to contribute to the reduction of greenhouse emissions (121).

2.21 Chitin and Chitosan

The natural biopolymer chitin and its deacetylated product chitosan are found abundantly in nature as structural building blocks and are used in all sectors of human activities like materials science, nutrition, health care, and energy (122).

These polymers are able to open opportunities for completely novel applications due to their exceptional properties in which an economic value is intrinsically entrapped. On a commercial scale, chitosan is mainly obtained from crustacean shells rather than from fungal and insect sources. Significant efforts have been devoted to commercialize chitosan extracted from fungal and insect sources to completely replace crustacean-derived chitosan. However, the traditional chitin extraction processes are ladened with many disadvantages.

The potential bioextraction of chitosan from fungal, insect, and crustacean, as well as its superior physicochemical properties, have been reviewed (122). The different aspects of fungal, insect, and crustacean chitosan extraction methods and various parameters having an effect on the yield of chitin and chitosan were detailed. In addition, the essential attributes of chitosan for high value-added applications in different fields have also been discussed.

2.22 Amylose Hybrids

Amylose and its hybrids are of basic interest for their use in biological applications. Methods have been developed for designing functional amylose hybrids (123). The phosphorylase-catalyzed polymerization method shows considerable promise as a tool for preparing diverse amylose hybrids.

The generation of amylose-functionalized materials from amylose primers has been reviewed (123).

Recently, advances have been made in the chemoenzymatic synthesis and characterization of amylose block polymers, amylose graft polymers, amylose-modified surfaces, hetero-oligosaccharides, and cellodextrin hybrids.

Many of these saccharides show clear opportunities for advances in biomaterials because of their biocompatibility and biodegradability. An appropriate choice of amylose and other materials enables the addition of further functionalities and tuning and control of the materials for specific applications. Newly developed amylose hybrids can help to promote the development of new generations of glyco materials.

Phosphorylase catalyzes the addition of glucose units from glucose monophosphate to the nonreducing end of an amylose primer, e.g., maltopentaose, in the absence of inorganic phosphate. The reducing end of the primer can be functionalized, and the resulting primer can be recognized for phosphorylase. Maltopentaose is shown in Figure 2.24.

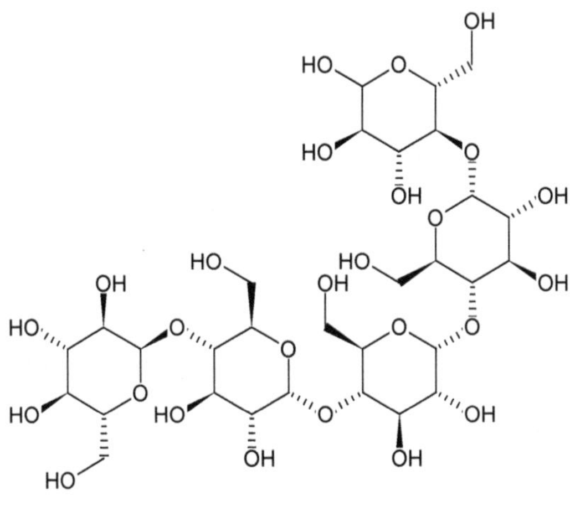

Figure 2.24 Maltopentaose.

This method for the enzymatic synthesis of amylose hybrids was first used in 1987 (124). Two amylose chain functionalized PEG were prepared. These were the first developed amylose-based copolymers. It was found that a bifunctional primer resulted in amylose block copolymers with a uniform molecular weight distribution of amylose. This was explained due to equal recognition of phosphorylases by both ends of the primer (123).

Amylose can wrap various hydrophobic polymers during vine-twining polymerization. A maltoheptaose-*b*-poly(*L*-lactic acid) (PLLA) was prepared via a copper-catalyzed 1,3-dipolar cycloaddition, which is a primer-guest molecule hybrid (125). The elongated amylose chains entrap other PLLA chains in their helical cavities during enzymatic polymerization, resulting in the formation of linear supramolecular polymers.

Also, a three-armed maltoheptaose PLLA conjugate was used as a primer for the construction of supramolecular polymers (125). After enzymatic polymerization, the resulting amylose PLLA formed an ionic gel in 1-butyl-3-methylimidazolium chloride, c.f. Figure 2.25. The gelation was probably caused by the formation of inclusion complexes between the branched polymers. A similar approach was used to show inclusion complex formation by an amylose poly-(tetrahydrofuran) conjugate (126).

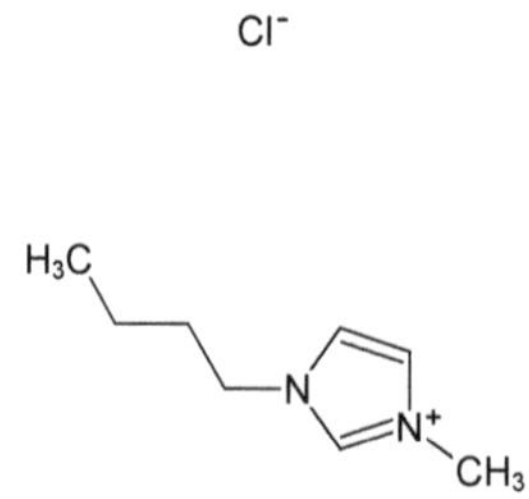

Figure 2.25 1-Butyl-3-methylimidazolium chloride.

Post-functionalizable two-dimensional crystalline cellulose nanosheets were prepared as follows (127):

An azide-containing cellulose oligomer was synthesized by reacting β-glycosyl azide as a primer and α-glucose monophosphate as a substrate in the presence of cellodextrin phosphorylase. The resulting cellulose oligomers aligned and formed nanosheets of an average thickness of 5.5 *nm*. It could be demonstrated that the nanosheet formed a cellulose II allomorph, which had an antiparallel saccharide arrangement against the sheet. The azide groups of cellodextrin were therefore located on the sheet surface. Such structures offer advantages such as the possibility of nanosheet post-functionalization (123).

A gene delivery system was developed based on enzymatically synthesized polysaccharides. The system consists of cycloamylose (CA) functionalized with cationic groups. CA, which is a macrocyclic polysaccharide consisting of (1,4)-glucose units, is produced through the treatment of linear amylose with 4-α-glucanotransferase.

A series of CA derivatives were reported that effectively delivered plasmid DNA (pDNA) (128), siRNA (129), and CpG DNA (130) *in vitro* and *in vivo* (131). Also, a novel gene delivery system was developed using a hexadecyl-group-bearing cationic CA and a phospholipid-degrading enzyme (132).

2.23 Poly(acryloyl hydrazide)

Poly(acryloyl hydrazide) is a versatile material for the preparation of functional polymers (133). Poly(acryloyl hydrazide) could be prepared from commercially available starting materials in a three-step synthesis on a large scale, in good yields and high purity.

Two synthetic strategies to achieve the target poly(acryloyl hydrazide) starting from commercially available *tert*-butyl-2-acryloylhydrazine-1-carboxylate, c.f. Figure 2.26, were investigated. In both cases, polymerization was employed because of its versatility for the preparation of acrylamide-based polymers (134–136).

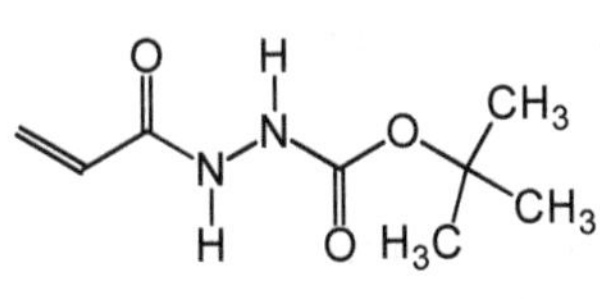

Figure 2.26 *tert*-Butyl-2-acryloylhydrazine-1-carboxylate.

The polymerization of a deprotected acryloyl hydrazide was investigated (133). Initial polymerization of this monomer was done with 2-((Ethylthio)carbonothioyl)thio-2-methylpropanoic acid, c.f. Figure 2.27, a chain transfer agent. Nuclear magnetic resonance spectroscopy studies indicated the formation of oligomeric compounds, probably resulting from the Michael addition of nitrogen in one hydrazide. So the polymerization of a protected monomer was attempted instead. Here, the reaction slowed down at higher conversions, suggesting termination.

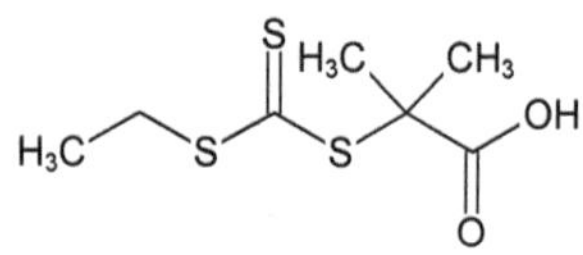

Figure 2.27 2-((Ethylthio)carbonothioyl)thio-2-methylpropanoic acid.

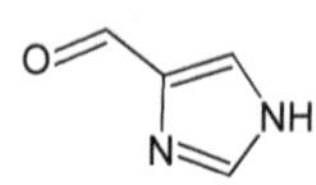

Figure 2.28 4-Imidazolecarboxaldehyde.

4-Imidazolecarboxaldehyde, c.f. Figure 2.28, was used as a model hydrophilic aldehyde to assess the conditions for the functionalization of the polymer scaffold (137). As reported, the coupling of 4-imidazolecarboxaldehyde with poly(acryloyl hydrazide) could be performed in an 100 *mM* acetic acid/D_2O buffer with varying amounts of aldehyde in relative short times of 1–4 *h*. The duration of the coupling reaction was not dependent on the number of equivalents added.

After deprotection under acidic conditions, it could be demonstrated that poly(acryloyl hydrazide) is a versatile reactive scaffold that can mediate the synthesis of polymers carrying a wide range of functionalities, including non-water-soluble and biologically similar aldehydes. The efficiency of the hydrazide-aldehyde coupling was modulated by tuning the reaction conditions, including the use of both aqueous and organic conditions, to yield polymers with a consistent degree of functionalization (133).

2.24 Redox-Active Tetrathiafulvalenes

Coordination polymers based on tetrathiafulvalene and its derivatives have been reviewed (138). In coordination polymers (CP-s), metal ions or metal-containing clusters act as nodes, and organic ligands act as spacers, both of which are linked via coordination bonds to form one-, two- or three-dimensional extended structures (139,140)

The design and study of multifunctional coordination polymers with multiple accessible properties has attracted attention, owing to their potential applications in a lot of areas, including sensing, solar energy harvesting, and energy storage, and others (141–143).

Functional coordination polymers can be rationally designed by using organic ligands with certain desirable properties such as redox activity. A promising candidate is tetrathiafulvalene, which is shown in Figure 2.29.

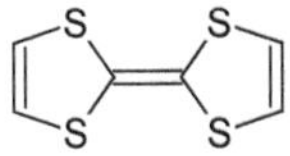

Figure 2.29 Tetrathiafulvalene.

Several tetrathiafulvalene derivates have been described, which are shown in Table 2.5. Some compounds are also shown in Figure 2.30.

The incorporation of bulkier benzoic acid pendant groups in the ligand tetrathiafulvalene tetrabenzoic acid, c.f. Figure 2.30, assisted in extending the length of the ligand to achieve larger pore coordination polymers. This ligand was synthesized through a palladium-catalyzed cross-coupling between tetrathiafulvalene and ethyl-4-bromobenzoate, c.f. Figure 2.31 (144).

2,9-Bis(4-pyridyl)tetrathiafulvalene is an analogue of 4,4'-bipyridine and could serve as a donor linker owing to its effective charge-transfer properties.

Also, flexible links have been introduced to tetrathiafulvalene to generate new functional ligands. The unsymmetrical tetrathiafulvalene derivative 2-[4,5-bis(methylsulfanyl)-1,3-dithiol-2-ylidene]-4,5-bis(2-cyanoethylsulfanyl)-1,3-dithiole, has been prepared by attach-

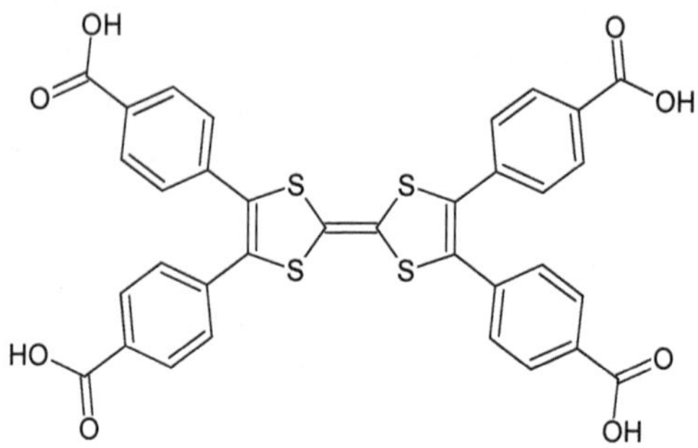

Tetrathiafulvalene tetrabenzoic acid

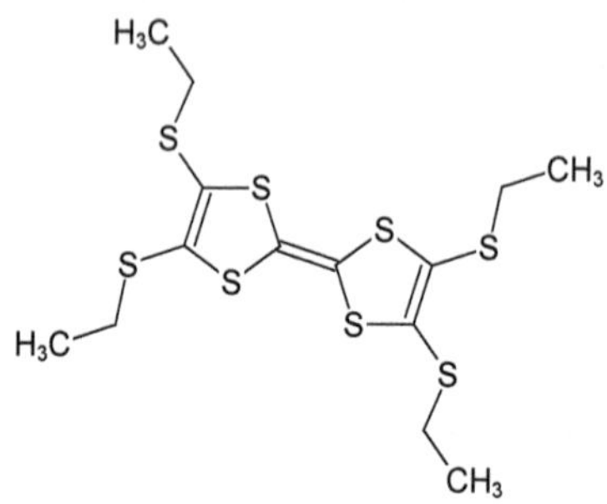

Tetrakis(ethylthio)tetrathiafulvalene

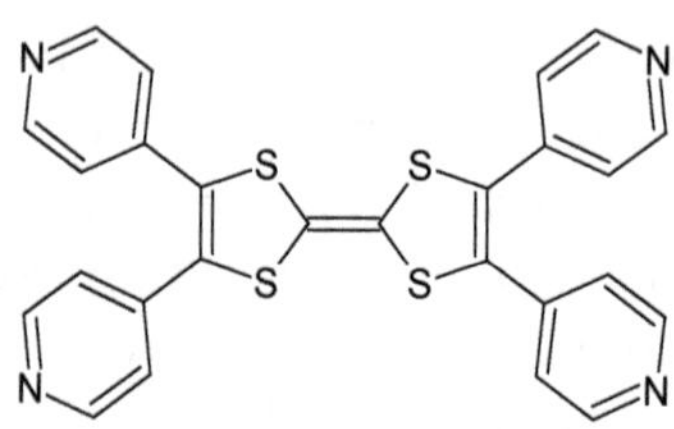

Tetra(4-pyridyl)-tetrathiafulvalene

Figure 2.30 Tetrathiafulvalene compounds.

Table 2.5 Tetrathiafulvalene derivates (138).

Compound
Tetrathiafulvalene
Tetrathiafulvalene tetracarboxylic acid
Tetrathiafulvalene tetrabenzoic acid
(4′-Pyridylmethylsulfanyl)-4′,5′-ethylenedithiotetrathiafulvalene
2,9-Bis(4-pyridyl)tetrathiafulvalene
Thenoyltrifluoroacetone
Tetra(4-pyridyl)-tetrathiafulvalene
Tetrakis(ethylthio)tetrathiafulvalene
Tetrakis(methylthio)tetrathiafulvalene
Tetrakis(propylthiothio)tetrathiafulvalene
Bis(ethylenedithio)tetrathiafulvalene
2,3-Bis(carboxyl)-6,7-bimethylthiotetrathiafulvalene

Tetrathiafulvalene Ethyl-4-bromobenzoate

Figure 2.31 Compounds for ligands (144).

ing the methylsulfanyl group at one end of tetrathiafulvalene and the cyanoethylsulfanyl group at the other end. This compound displays a unique coordination flexibility (138).

2.25 Sequence-Controlled Polymers

A method for preparing sequence-controlled polymers has been demonstrated by using a latent monomer, a furan-protected maleimide, c.f. Figure 2.32 (145). At 110°C, the furan-protected maleimide is deprotected by a retro Diels-Alder reaction, and the released maleimide is immediately involved in the cross-polymerization with styrene to deliver heterosegments. At 40°C the retro Diels-Alder reaction does not proceed, therefore homo-PS segments are produced.

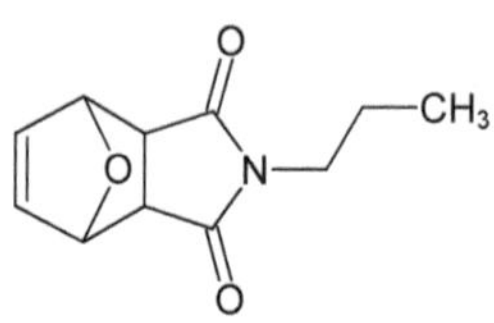

Figure 2.32 Furan-protected maleimide.

By implementing programmable temperature changes during polymerization of styrene and furan-protected maleimide, living polymers with a tailored sequence are created. A ternary copolymerization can produce complex sequences as designed.

Alkynyl-functionalized furan-protected maleimide, used as a latent monomer, leads to the desirable placement of functional groups along the polymer chain (145).

2.25.1 *Sequential Thiol-ene and Amino-yne Click Reactions*

Sequence-controlled polymers have been prepared by the combination of thiol-ene click reaction and amino-yne click reaction (146). Due to the high selectivity of the amine unit and thiol unit toward thiolactone, methacrylate, propiolate, etc., the sequence of the resulting polymers can be easily controlled by the sequence of monomer addition. In addition, the thiol-ene click reaction and the amino-yne click reaction can be carried out under mild conditions without the need of any catalyst.

Various kinds of sequence-controlled copolymers have been prepared by the combination of the thiol-ene click reaction and the amino-yne click reaction in one pot. The synthesis of *N*-acetylhomocysteine thiolactone, which is used as monomer, was done as follows (146):

Preparation 2–1: First, 3.07 *g* (20 *mmol*) *D,L*-homocysteine thiolactone hydrochloride was mixed with 9.70 *g* (96 *mmol*) triethylamine in 50 *ml* dichloromethane in an ice bath. Subsequently a suspension was formed. Then 2.36 *g* (30 *mmol*) acetyl chloride was added dropwise over 30 *min*. This solution was stirred at room temperature for an additional 4 *h*. The reaction mixture was diluted with 20 *ml* dichloromethane, filtered, washed with brine, and extracted with dichloromethane. The organic phase was dried with anhydrous Na_2SO_4. The resulting solution was subjected to flash silica chromatography with ethyl acetate as eluent. A white powder was obtained after evaporation of the solvent with a yield of 75%.

The preparation of a polymer using *N*-acetylhomocysteine thiolactone was done as follows (146):

Preparation 2–2: First, 88 *mg* (0.1 *mmol*) *N*-ethyl ethylene diamine and 99 *mg* (0.05 *mmol*) ethylene dimethacrylate were dissolved in 1.0 *ml* of *N,N*-dimethylformamide under argon atmosphere and stirred. Then 160 *mg* (0.1 *mmol*) *N*-acetylhomocysteine thiolactone was added, and the reaction mixture was left at 50°C for 3 *h*. Subsequently, 111 *mg* (0.05 *mmol*) 1,6-hexanediol dipropiolate, c.f. Figure 2.33, was added into the reaction mixture, and the solution was stirred at room temperature for an additional 3 *h*. The polymer was obtained as yellow oil with a yield of 82% by precipitating into diethyl ether.

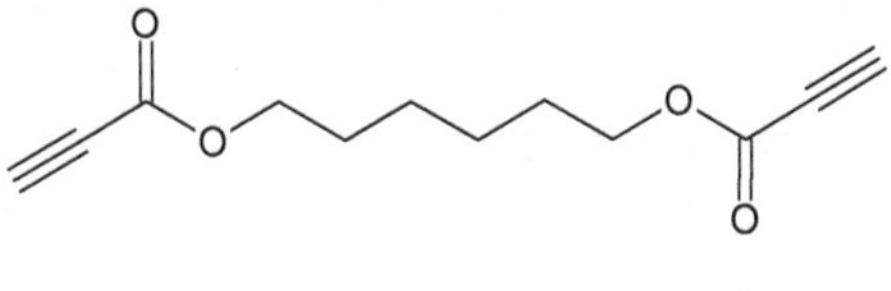

Figure 2.33 1,6-Hexanediol dipropiolate.

Some chemicals mentioned above are shown in Figure 2.34. The formation of polymers with high molecular weight and controlled sequence structure could be confirmed (146).

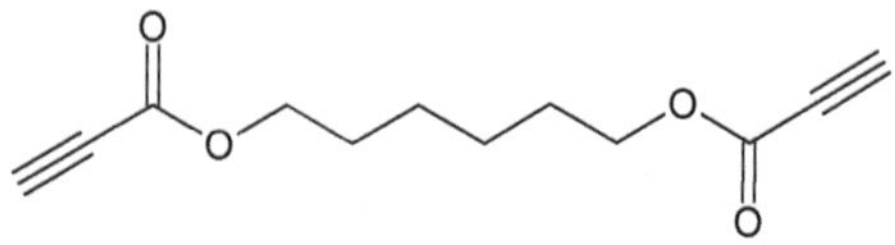

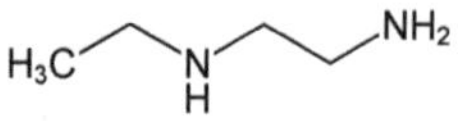

1,6-Hexanediol dipropiolate *N*-Ethyl ethylene diamine

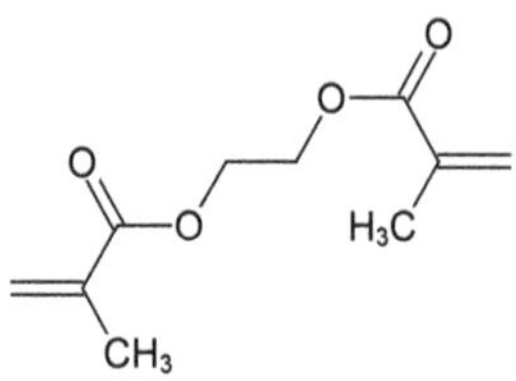

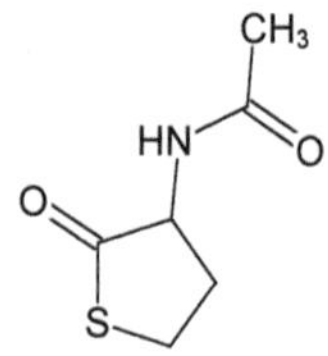

Ethylene dimethacrylate *N*-Acetylhomocysteine thiolactone

Figure 2.34 Materials for thiol-ene and amino-yne click reactions.

2.26 Oligomeric Silsesquioxane

The continuous demand for novel hybrid materials in specific technological applications inspires researchers to develop new synthetic strategies in a modular and efficient way. In recent years, extensive efforts have been devoted to using polyhedral oligomeric silsesquioxane (POSS) to construct multifunctional nanohybrids and nanocomposites with tunable hierarchical structures and unparalleled properties (147). A POSS material is shown in Figure 2.35.

The shape-persistent nanostructure and diverse surface chemistry make these nanocaged materials ideal building blocks for such purposes. Functionalization of POSS cages were further facilitated by the introduction of click chemistry at the beginning of this century. Click reactions include several kinds of selective and orthogonal chemical ligations with high efficiency under mild reaction conditions. The concept has generated real stimulus not only in elegantly preparing materials of choice, but in making the leap from laboratory to industrial scale-up of POSS-based hybrid materials as well (147).

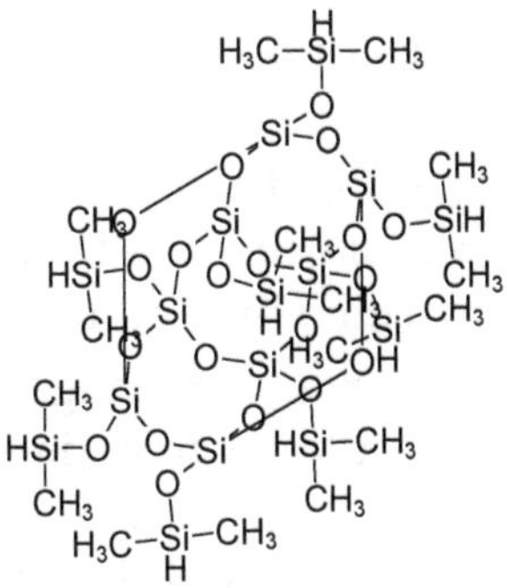

Figure 2.35 Octakis(dimethylsiloxy)-T8-silsesquioxane.

2.27 Proteins

A series of methods have been outlined for attaching functional polymers to proteins in a monograph (148). Polymers with a good control over structure, functionality, and composition can be created using reversible addition-fragmentation chain transfer (RAFT) polymerization. These polymers can be covalently linked to enzymes and proteins using either the *grafting-to* approach, where a preformed polymer is attached to the protein surface, or the *grafting-from* approach, where the polymer is grown from the protein surface.

Methods for grafting-to, or attaching the RAFT chain transfer agent to the protein surface, include the commonly used carbodiimide/activated ester coupling. Methods were also outlined to graft-from the surface of the protein using RAFT polymerization.

In addition, it is possible to site specifically introduce a reactive azide group to the protein surface using enzymatic ligation as a posttranslational modification. This reactive azide group can be conjugated to an alkyne-containing polymer using highly efficient click chemistry. These methods can produce protein polymer conjugates with various architectures and functionalities (148).

The structure and the activity of proteins serve as important applications of functional polymeric materials. Several attempts to mimic the various structural and functional traits of proteins have been tried using the techniques of modern polymer chemistry.

The properties and utility of proteins in applications, such as catalysis and molecular recognition, have been emulated in the laboratory.

The control of the sequence of repeating units in polymers with the precision of biomacromolecules is a frontier in polymer synthesis. This topic has been covered by a number of reviews (149–151)

Appropriate sequences for a wide range of interesting material targets, ranging from molecules to materials surfaces and internal interfaces, can be selected via combinatorial means, and sequence specificities within the resulting peptide-target interactions can be routinely investigated (152).

Based on this understanding, macromolecular sciences can define new polymer structures that meet the required functionalities or functional sequences with fully synthetic, nonpeptidic precision polymers to endeavor an information-based design of next-generation, purpose-adapted macromolecules.

2.27.1 *Iterative Techniques*

Iterative techniques typically utilize difunctional monomers with one protected functional group. After reaction of the unprotected functionality, the protecting group is removed, thereby exposing a new site for monomer addition.

An iterative strategy has been developed (153) that is based on the Passerini three-component reaction (154, 155). In such a system, an isocyanide bearing a benzyl ester is reacted with both an aldehyde and a carboxylic acid to form a depsipeptide. Then, the benzyl ester is cleaved, exposing a carboxylic acid, which is subsequently reacted with the same isocyanide and another aldehyde. The different functionalities of the aldehyde monomers become pendant groups, and are placed in a specific order along the growing polymer chain.

2.27.2 *Sequencing Techniques*

Natural polymers can be sequenced using templates that migrate around the growing chain end. A synthetic system for sequence control using a migrating template has been described (156). In this system, a controlled radical polymerization initiator and a vinyl monomer functionality are attached to orthogonally cleavable points on a single molecule. After a single monomer insertion to close the macrocycle, one of the functionalities is cleaved to open the macrocycle. Another vinyl monomer is attached to the cleaved functionality and the macrocycle is closed again, adding another monomer unit.

The repetitive reaction, purification, and deprotection of the iterative synthetic approach can create polymers that are aesthetically similar to biological polymers, but the polymerization mechanism could not be more different. The ability to template truly sequenced high molecular weight polymers with non-natural backbones, although currently a technological impossibility, remains an important challenge for research (157).

2.27.3 *Ring-Opening Polymerization*

In order to create a longer sequence of repeat units, defined segments can be programmed into a macrocycle and then ring-opening polymerized, creating sequenced ring-opening metathesis polymerization (ROMP) polymers. When long sequences are desired, the ring will be unstrained, making traditional ROMP techniques ineffective for polymerization.

An entropy-driven ROMP has been assessed as a route to create sequence-controlled copolymers of lactate, glycolate, and ε-caprolactate (158). Entropy-driven ROMP is not a living polymerization technique, but is still capable of molecular weight control as opposed to the closely related acyclic diene metathesis polymerization, because of the ring-chain equilibrium. Sequence-controlled entropy-driven ROMP polymers were made with molecular weights up to 78 *k Dalton* and with 86% conversion.

2.27.4 *Foldable Polymers*

A strategy for preparing foldable linear polymer chains was demonstrated, using the control of monomer sequence (159,160). The controlled addition of discrete amounts of functionalized maleimide at precise times during the synthesis enabled the formation of PS chains that contained positionable reactive alkyne groups that could then react intramolecularly to fold the chain. A variety of folded polymers were obtained, including tadpole (P-shaped), pseudocyclic (Q-shaped), bicyclic (8-shaped), and knotted (α-shaped) macromolecular shapes.

2.27.5 *Synthetic Enzymes*

When designing a synthetic enzyme mimic, two considerations are of importance. The first consideration is the catalytic center and environment directed attached to the catalytically active center. Many active site functions are predicated around this primary coordination sphere.

The second consideration is the secondary coordination sphere provided by the macromolecular environment that surrounds these sites. To better understand and utilize the structures and functions of proteins, scientists have sought to design synthetic macromolecular models in an attempt to achieve the same complexity and utility found in nature (157).

Using single-chain nanoparticles (161–163) and supramolecular chemistries, a series of polymer scaffolds were generated amendable to site-isolated catalysis (164–166).

A [Fe-Fe] hydrogenase mimic has been synthesized that could be attached to a single-chain nanoparticle, thus generating the first single-site bioinspired single-chain nanoparticle (167). This bioinspired mimic was made using a copolymer of methyl methacrylate (MMA) and an anthracene-derived methacrylate.

The [Fe-Fe] cluster was covalently attached via a thiol-ene click chemistry after aminolysis of the trithiocarbonate group. Upon photodimerization, the polymer crosslinked to generate a hydrophobic environment around a single [Fe-Fe] synthetic model site. This system provides scaffolds that possess a site-isolated catalytic center (167).

A porphyrin-cored polymer nanoparticle that was designed using a tetrafunctionalized porphyrin macroinitiator and MMA and 9-anthracenylmethyl methacrylate, c.f. Figure 2.36, as comonomers (168).

The resultant nanoparticle was probed as a heme model mimic by exposing the system to reductants and exogenous ligands. The inherent tunability of these systems allows for probing of different macromolecular microenvironments, such as hydrophilic, hy-

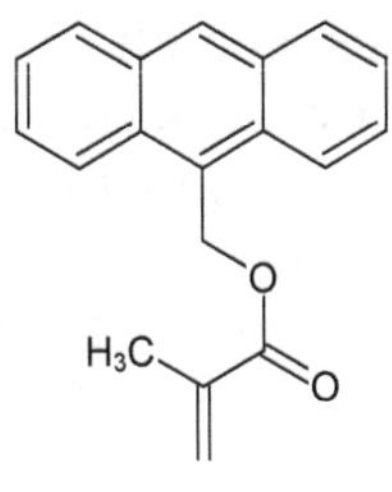

Figure 2.36 9-Anthracenylmethyl methacrylate.

drophobic, and hydrogen-bonding-rich, to investigate the reactivity of the model heme iron center toward relevant biological ligands (168).

2.27.6 *Synthetic Membrane Proteins*

Proton conducting membranes are found in almost all forms of life. The interactions between the membranes and membrane proteins are crucial for efficient proton and electron transfers (169,170).

Photosynthesis utilizes these transfers to complete its cycle. Recently, Sun *et al.* designed a suite of phosphonated peptoid diblock copolymers using poly(*N*-(2-ethyl)hexylglycine) and poly(*N*-(phosphonomethyl)glycine) to determine how the morphology of hydrated polymers affected the proton conductivity of the material (171). These monomers are shown in Figure 2.37.

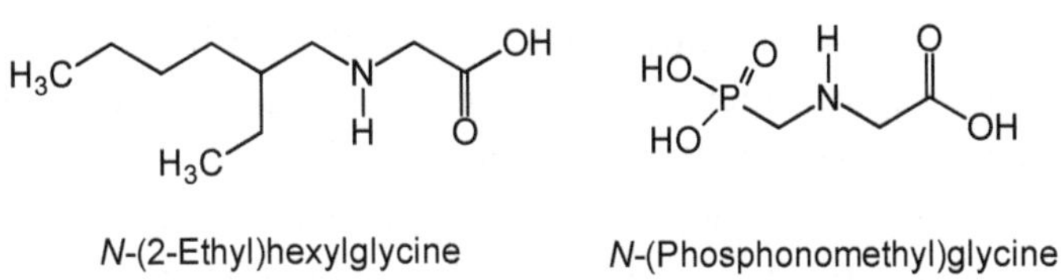

Figure 2.37 Monomers for phosphonated peptoid diblock copolymers (171).

It was found that the morphology played a role in dictating whether or not the material would be insulating or conducting. The disordered polymer type yielded insulating properties with a conductivity of < 10^{-7} $S\,cm^{-1}$. Conversely, the ordered polymers showed conductivities as high as 0.008 $S\,cm^{-1}$. It was also found that a more symmetric block copolymer with higher molecular weight showed a greater conductivity. So, this study provided a valuable foundation for using peptoid block copolymers as phosphonated polymer electrolytes in electrochemistry and proton conduction (157).

2.28 Functional Polymer Microspheres

Functional particles can be prepared directly by heterogeneous polymerization such as emulsion polymerization and dispersion polymerization (172). The modification of certain existing particles is another method to prepare functional particles.

The features of nanoparticles or microparticles, such as large specific surface area, high mobility, easy recovery from the dispersion and reversible dispersibility, etc., can be utilized for the exhibition of desired functions.

Medical and biochemical applications of such particles, including absorbents, latex diagnostics, affinity bioseparators and drug and enzyme carriers, are the most practical ones. Also, optical and opto-electrical functions of such particles are attracting attention.

Some particles may exhibit a unique rheological behavior under special conditions and thus can expand their applications (172).

2.28.1 *Anisotropic Nonspherical Functional Polymeric Particles*

A feasible and versatile route for a large-scale synthesis of monodisperse nonspherical functional polymer particles by one-pot dispersion polymerization has been demonstrated (173).

The particle morphology could be precisely tuned by varying the divinylbenzene (DVB) concentration, the start feeding time of DVB, the total feeding time of DVB, the styrene mass ratio of stage 1/stage 2, and the solvent polarity. Sphere-like, polyhedron-like, and red-blood-cell-like particles with dimples or smooth surface were obtained.

In addition, the formation mechanism of these nonspherical particles was attributed to the phase separation induced by the uneven distribution of the crosslinked network. Because of the incompatibility between a styrene homopolymer chain and the crosslinked network, the styrene homopolymer chain was forced to move to those zones where blending was allowed. The particle surface was forced to distort, leading to the formation of particles with various morphologies (173).

2.28.2 *Ion Exchange Resins*

Ion exchange resins have been widely used as acid or base catalysts. Polymer-supported transition metal catalysts and phase transfer catalysts are also commonly used particles. They are usually larger than 10 μm. Ion exchange materials are detailed in Section 3.13 of this monograph.

2.28.3 *Functional Aromatic Poly(amide)s*

The state-of-the-art of aromatic PAs, wholly aromatic PAs or aramids has been reviewed (174). These polymers belong to the family of high-performance materials because of their exceptional thermal and mechanical behavior. These materials have been transformed into fibers mainly for production of advanced composites, paper, and cut and fire protective garments.

The preparation of aromatic PA dendrimers has been reported. A divergent approach was used, up to the sixth generation with a polyhedral oligomeric silsesquioxane (POSS) (175). An eight-functional core was prepared using octa(*n*-propylamine)-POSS and an AB_2 building block using 3,5-bis(trifluoroacetamido)benzoic acid, c.f. Figure 2.38, and deprotected with hydrazine to yield the G_1 dendrimer. The rest of the dendrimers that were synthesized in a similar way, were obtained in high yields without tedious purification. These organic-inorganic hybrid nanoparticles have a potential application in nanoscience as fluorescent nanofillers with controlled functionality, shape and size.

A one-pot multistep synthesis was performed by preparing dendritic aromatic PA dendrons on a glycine-modified Wang resin as the solid support (176). 3,5-Bis(4-(9-fluorenylmethyloxycarbonyl)-aminophenoxy)benzoic acid and (2,3-dihydro-2-thioxo-3-benzoxazolyl)phosphonic acid diphenyl ester were used as building block

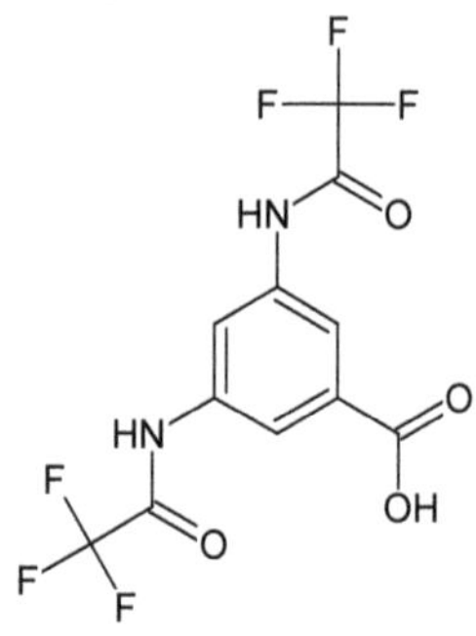

Figure 2.38 3,5-Bis(trifluoroacetamido)benzoic acid.

and as condensation agent, respectively. Any unreacted chemicals could be removed by washing the resin with solvents, as the propagating dendrons are attached to the insoluble support.

2.28.3.1 *Hyperbranched Aromatic Poly(amide)s*

Aromatic and semiaromatic amine-terminated hyperbranched PAs and carboxylic acid-terminated hyperbranched PA esters were synthesized (177). An aromatic triamine, 1,3,5-tris(4′-aminophenylcarbamoyl)benzene, was synthesized. The synthesis is shown in Figure 2.39.

The polycondensation reaction of this compound with terephthaloyl chloride, isophthaloyl chloride, sebacoyl chloride, and adipoyl chloride resulted in the formation of four hyperbranched PAs. The compounds used here are shown in Figure 2.40.

These thermally stable amorphous compounds were soluble in polar aprotic solvents at room temperature with glass transition temperatures between 138°C and 198°C (177).

Fluorescent porous hyperbranched aramids with different terminal functional groups have been synthesized (178). 1,3,5-Tri(4-carboxyl phenyl)benzene and *p*-phenylenediamine, c.f. Figure 2.41, were used to prepare PAs with strong blue fluorescence with potential applications in storage, separation, catalysis and light emission.

Other applications of hyperbranched PAs are their use in reverse osmosis membranes (179), in applications related to their magnetic properties due to the complexation of hyperbranched PAs with a

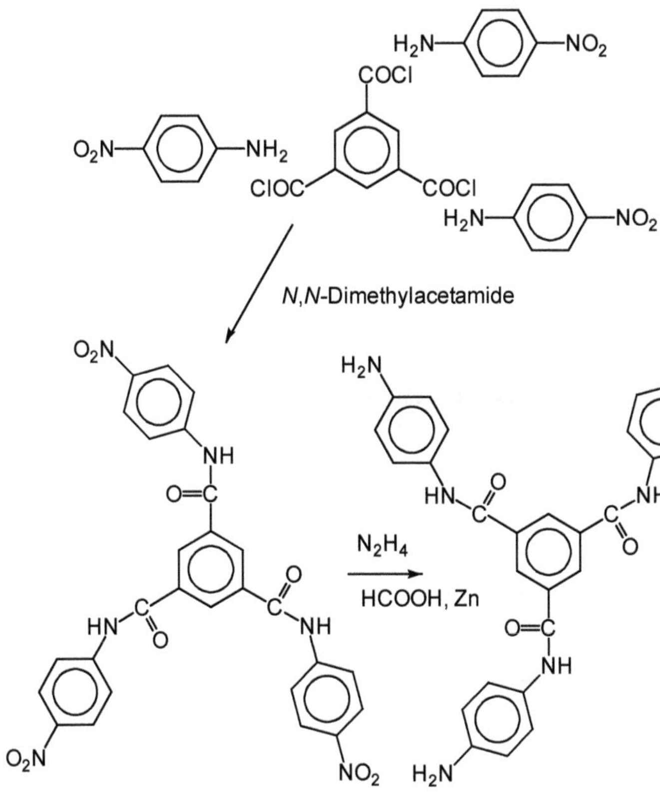

Figure 2.39 Synthesis of 1,3,5-tris(4′-aminophenylcarbamoyl)benzene (177).

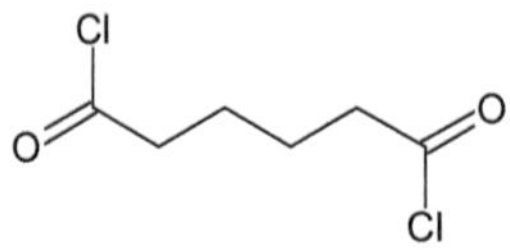

Adipoyl chloride

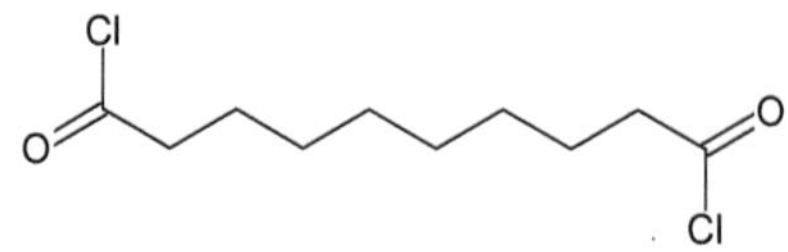

Sebacoyl chloride

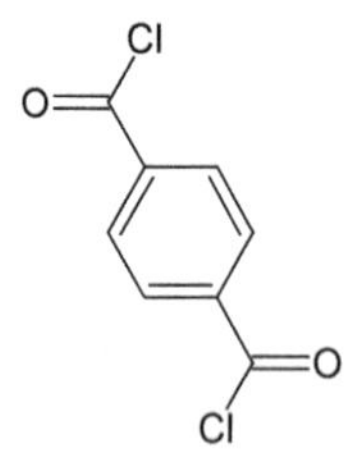

Terephthaloyl chloride

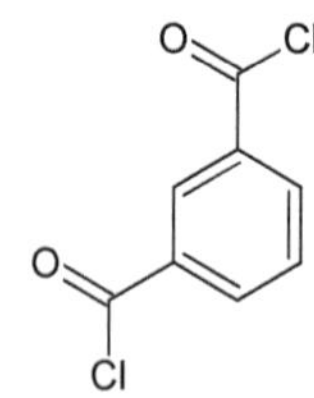

Isophthaloyl chloride

Figure 2.40 Acid chlorides.

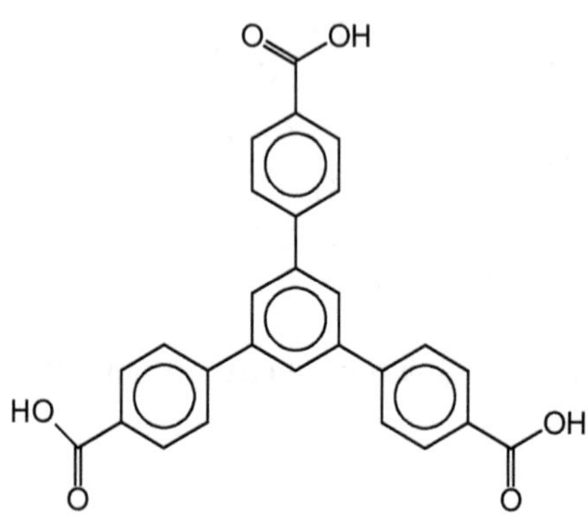

1,3,5-Tri(4-carboxyl phenyl)benzene

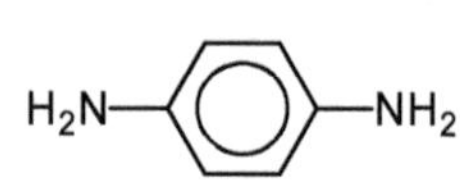

p-Phenylenediamine

Figure 2.41 Materials for aramids.

metal (180), to their electrochromic properties after the introduction of electroactive units (181), or the improvement of thermal properties of composites grafted with hyperbranched PAs (182, 183)

2.28.4 *Enzyme Immobilization*

Other particulate catalysts are microspheres immobilizing functional biopolymers, especially enzymes and their analogues. The advantage of enzyme immobilization is easy recovery and repetitive use of enzymes. Sometimes, the immobilization improves the durability of the enzyme (172).

Enzyme immobilization can be done by several methods (184):

1. Adsorption,
2. Entrapment, and
3. Chemical binding.

For example (184), peroxidase and antitransferrin of anti-α-fetoprotein were co-immobilized to tresyl activated 40 *nm* silica particles and the particle conjugates used in heterogeneous enzyme immunoassays for the evaluation of transferrin and α-fetoprotein within the concentration range of 0.25–1000 $ng\,ml^{-1}$.

In general, adsorption and entrapment result in less stress and less activity loss to the enzyme, but may give uncertain immobilization compared to chemical binding (172).

For chemical binding, a suitable functional group for binding and the binding conditions must be carefully chosen to prepare hybrid particles with high activity. Functional groups for binding can be carboxyl, amine, hydroxyl, thiol, epoxy groups, etc. These functional groups are set on the particle surface by copolymerization of the core-forming monomer with functional monomers, such as (meth)acrylic acid, glycidyl methacrylate, and active ester monomer such as nitrophenyl acrylate. These compounds are shown in Figure 2.42.

In order to avoid a steric hindrance to the enzymic reaction, enzymes are sometimes immobilized to particles via a so-called spacer such as an oligo ethylene glycol or a peptide. A thermosensitive polymer, poly(*N*-isopropylacrylamide), was used as a spacer for trypsin (185). Trypsin is shown in Figure 2.43.

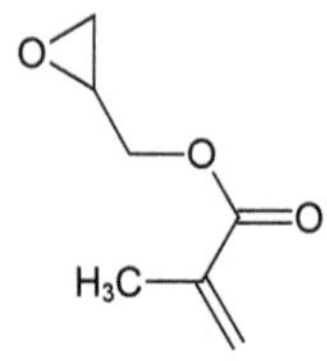

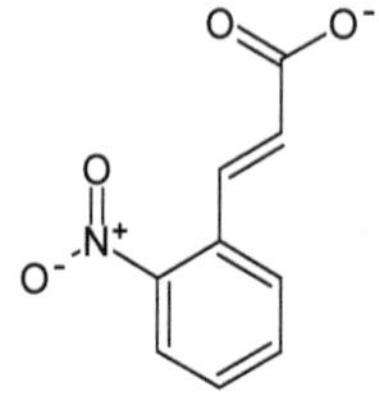

Glycidyl methacrylate Nitrophenyl acrylate

Figure 2.42 Functional monomers.

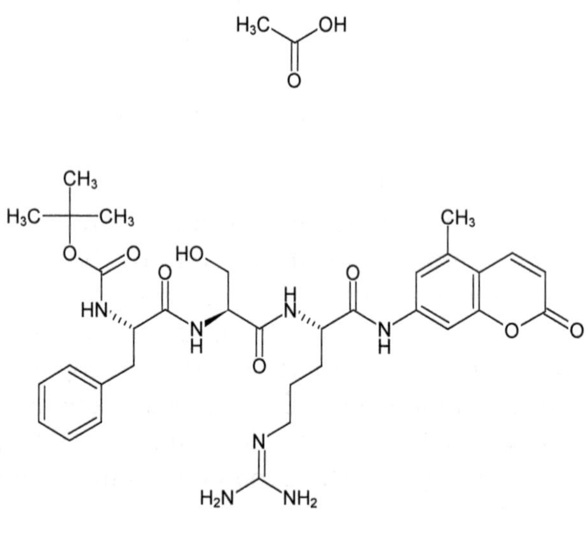

Figure 2.43 Trypsin.

Poly(*N*-isopropylacrylamide) has a transition temperature (or the lower critical solution temperature) at 32°C and, below this temperature, the gel is swollen with water. Above the lower critical solution temperature it releases water.

When used as a spacer for trypsin, according to differential scanning calorimetry studies, the transition temperature of poly(*N*-isopropylacrylamide), to the chain end at which trypsin was bound, changed to 38°C. Because the poly(*N*-isopropylacrylamide) chains have free ends that coexist with trypsin-carrying poly(*N*-isopropylacrylamide) on the particles, the particles responded to temperature by two steps.

Above 32°C, poly(*N*-isopropylacrylamide) chains having free ends collapsed but trypsin-carrying poly(*N*-isopropylacrylamide) kept hydrated up to 38°C. Below 32°C, immobilized trypsin is surrounded by free chain-end poly(*N*-isopropylacrylamide), blocked from the substrate in the medium, and has a low relative activity. In contrast, above 32°C, the collapsed free chain-end poly(*N*-isopropylacrylamide) does not interfere with the access of substrate to enzyme anymore, and the enzyme gets more of a chance to react with the substrate and has a higher relative activity. The temperature-dependent activity was observed only in the case using bulky substrates, so this system may be applicable to selective enzyme reaction as a function of the molecular size of substrates (172).

2.29 Functional Biopolymers

2.29.1 *Saccharide-Based Helical Polymers*

Polysaccharides, such as cellulose and amylose, are abundant and optically active carbohydrate-based resources (186). Cellulose and amylose contain three hydroxyl groups in their glucose repeating units, which can be easily derivatized to afford desired functional pendants through polymer reactions with suitable reagents.

Suitably modified derivatives, such as benzoate and phenylcarbamate derivatives, exhibit excellent resolution abilities for a wide variety of chiral compounds when applied to chiral stationary phases for high-performance LC. Despite the established application of polysaccharides in chiral stationary phase materials, their use in other chiral functional materials only has been limited.

The functionalization of cellulose and amylose to yield novel chiral functions other than those in chiral stationary phases has been reviewed (186). Also, the use of these materials for various applications, such as asymmetric organocatalysts, chiral auxiliaries and chiral fluorescent sensors, has been detailed. The synthesis of a saccharide-containing polymer via the polymerization of a glucose-based monomer and its application as a circularly polarized luminescent material were also described (186).

Various monosaccharide-based chiral auxiliaries have been developed for use in asymmetric reactions, including 1,3-dipolar cycloadditions, 1,4-conjugate additions, Diels-Alder cycloadditions, aza-Friedel-Crafts reactions, alkylations, cyclopropanations and Strecker- and Mannich-type reactions (187–194).

The ready availability of polysaccharides and their potential for use as chiral functional materials make them intriguing scaffolds for chiral auxiliaries. The first example of a polysaccharide-based chiral auxiliary was reported in 2014 (186,195).

Here, cellulose and amylose derivatives bearing bromobenzoate pendants were synthesized as chiral auxiliaries to create optically active biaryl compounds through Suzuki-Miyaura cross-coupling (196) with naphthalen-1-ylboronic acid, c.f. Figure 2.44.

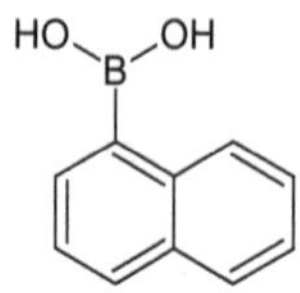

Figure 2.44 Naphthalen-1-ylboronic acid.

2.29.2 *Methacrylated Epoxidized Sucrose Soyate*

Previous studies showed that methacrylated epoxidized sucrose soyate can yield thermosets with a high glass transition temperature and good mechanical properties, but had a high viscosity of the resin due to hydrogen bonding of the hydroxyl groups (197). Further functionalization to yield dimethacrylated epoxidized sucrose soyate resulted in a reduced resin viscosity, but some of the thermosets were found to be brittle.

A study was done to evaluate the impact of modification of epoxidized sucrose soyate using a combination of methacrylate and inert esters (acetate propionate, butyrate) on the properties of the resins and thermosets produced from these compounds.

In order to maintain low resin viscosity and to improve the thermoset ductility, the replacement of some methacrylate groups with various other ester groups was explored (197). The synthesis of these resins was carried out in a one-pot process involving the sequential slow addition of anhydrides of the acids mixed prior to addition. The so synthesized resins were characterized for their viscosity with and without styrene as diluent.

Biobased resins with 30% styrene showed much lower viscosities compared to commercial resins containing higher amounts of styrene of 33% and 45%. The produced thermoset compositions had an improved flexibility and toughness with only a slight reduction in their glass transition temperature. The inclusion of non-functional esters in the resin structure acted as an internal plasticizer for the polymer composition (197).

2.29.3 *Mussel-Inspired Fabrication of Functional Materials*

Mussel-inspired chemistry has recently emerged as one of the most important and interesting surface modification methods owing to its gentle experiment conditions, high modification efficiency and universality (198).

Mussel-inspired chemistry mainly relies on the adhesion of dopamine on various materials and surfaces. In recent years, great efforts have been devoted to understanding the adhesion mechanism of mussels and extending its applications in different fields.

The recent development of mussel-inspired fabrication of functional materials for environmental applications has been presented (198). The surface modification strategies based on mussel-inspired chemistry were outlined in the first part. The environmental applications (e.g., oil/water separation, environmental adsorption and catalysts) based on these functional materials were also highlighted. Although many advances have been achieved, there still is plenty of room for further development of mussel-inspired surface chemistry for functional materials and their environmental applications (198).

2.29.4 *Conversion of Plant Biomass to Furan Derivatives*

5-Hydroxymethylfurfural is an important versatile reagent, a so-called platform chemical that can be produced from plant biomass compounds, i.e., hexose carbohydrates and lignocellulose.

In the near future, 5-hydroxymethylfurfural and its derivatives could become an alternative feedstock for the chemical industry and replace, to a great extent, non-renewable sources of hydrocarbons, i.e., oil, natural gas, and coal. A review has been presented that analyzes the recent advances in the synthesis of 5-hydroxymethylfurfural from plant feedstocks and considers the prospects for the use of 5-hydroxymethylfurfural in the production of monomers and polymers, porous carbon materials, engine fuels, solvents, pharmaceuticals, pesticides and chemicals (199).

The key synthetic pathways to 5-hydroxymethylfurfural are based on the acid- or metal-catalyzed dehydration of hexoses. Monosaccharides, disaccharides and polysaccharides, i.e., fructose, glucose, saccharose, inulin, starch, cellulose, etc., are used as the hexose feedstock. Methods have been developed for the conversion of fructose, glucose, saccharose and inulin in the presence of ethanol using various bifunctional catalysts, Brønsted and Lewis acids, functionalized ionic liquids, heteropolyacids, zeolites, graphene oxide and others.

The dehydration of fructose in isopropyl alcohol in the presence of heterogeneous catalysts based on mesoporous carbons functionalized on the surface by arylsulfonyl groups may give 5-hydroxymethylfurfural in a yield of up to 91%. However, the catalyst activity decreases after regeneration. High yields of 5-hydroxymethylfurfural up to 100% are obtained if the fructose dehydration is conducted in ionic liquids. As ionic liquids, organic salts are used, most often *N*-butyl-*N'*-methyl- and *N*-butyl-*N'*-ethylimidazolium chlorides and fluoroborates.

The most important 5-hydroxymethylfurfural derivatives considered include 2,5-furandicarboxylic acid, 2,5-diformylfuran, 2,5-bis(hydroxymethyl)furan, 2,5-bis(aminomethyl)furan, 2,5-dimethylfuran, 2,5-dimethyltetrahydrofuran, 2,5-bis(methoxymethyl)furan, and 5-ethoxymethylfurfural. These compounds are shown in Figure 2.45. In the near future, a significant extension of the 5-hydroxymethylfurfural application is expected, and this platform chemical may be considered to be a major source of carbon and hydrogen for the chemistry of the 21st century (199).

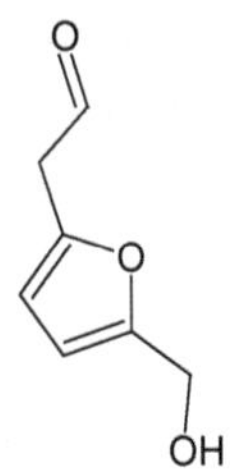

5-Hydroxymethylfurfural

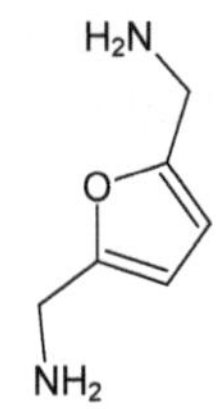

2,5-Bis(aminomethyl)furan

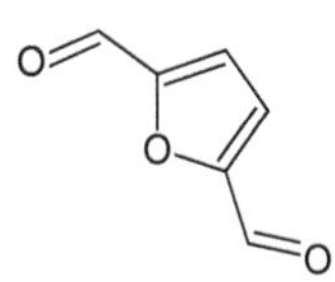

2,5-Diformylfuran

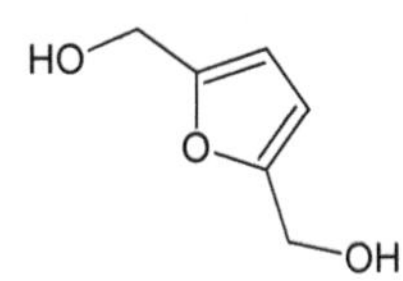

2,5-Bis(hydroxymethyl)furan

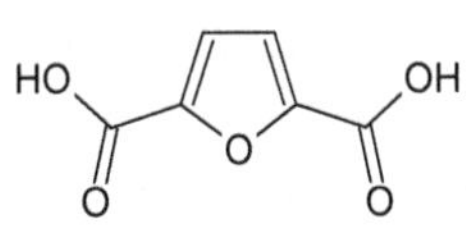

2,5-Furandicarboxylic acid

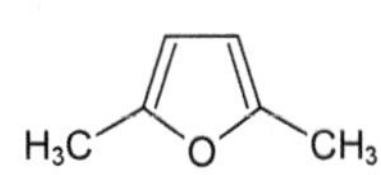

2,5-Dimethylfuran

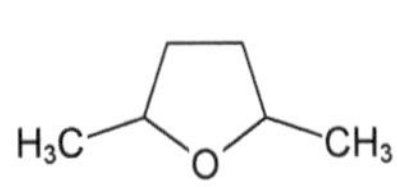

2,5-Dimethyltetrahydrofuran

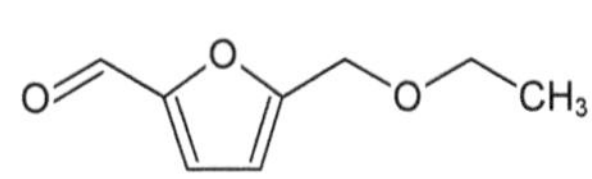

5-Ethoxymethylfurfural

Figure 2.45 5-Hydroxymethylfurfural derivatives.

2,5-Bis(alkoxymethyl)furans are considered efficient diesel fuel components for increasing the cetane number. The condensation of 5-hydroxymethylfurfural with carbonyl compounds followed by hydrogenation can serve as a source of normal hydrocarbons, which can be used as diesel fuel or aviation kerosene components.

Since 5-hydroxymethylfurfural is a polyfunctional reagent, it is also used for the preparation of biologically active compounds such as pharmaceuticals, pesticides, and nutrient additives.

5-Aminolevulinic acid and its derivatives are herbicides. Methods for the synthesis of 5-aminolevulinic acid hydrochloride based on 5-hydroxymethylfurfural and 5-chloromethylfurfural have been developed, c.f. Figure 2.46.

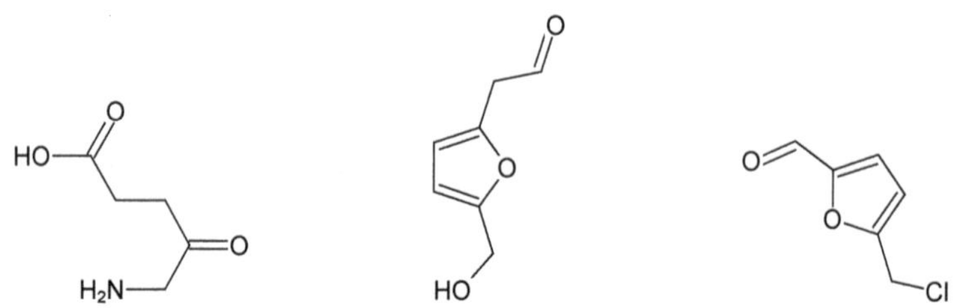

5-Aminolevulinic acid 5-Hydroxymethylfurfural 5-Chloromethylfurfural

Figure 2.46 Compounds for syntheses of 5-aminolevulinic acid.

Also, a synthetic route to the insecticide Prothrin, c.f. Figure 2.47, based on 5-chloromethylfurfural has been elaborated. The route includes the reaction of 5-chloromethylfurfural dibutyl acetal with trimethylsilylacetylene, the reduction of the aldehyde group and acylation with chrysanthemic acid chloride (199).

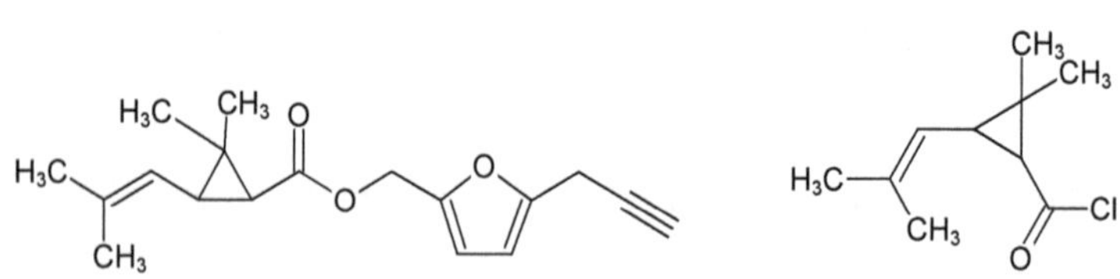

Prothrin Chrysanthemic acid chloride

Figure 2.47 Prothrin and Chrysanthemic acid chloride.

2.29.5 *Poly(deoxyribonucleotide) Analogues*

The use of deoxynucleoside building blocks has been shown. These building blocks are bearing non-natural bases. So, an attempt was made to develop a synthetic methodology that allows for the construction of high molecular weight deoxynucleotide polymers (200).

A six-membered cyclic phosphoester ring-opening polymerization was demonstrated by the initial preparation of polyphosphoesters, consisting of butenyl-functionalized deoxyribonucleoside repeat units connected via 3′,5′-backbone linkages.

A thymidine-derived bicyclic monomer, 3′,5′-cyclic 3-(3-butenyl) thymidine ethyl phosphate, was synthesized in two steps directly from thymidine via butenylation and diastereoselective cyclization promoted by *N*,*N*-dimethyl-4-aminopyridine.

Computational modeling of the six-membered 3′,5′-cyclic phosphoester ring derived from deoxyribose indicated strain energies at least 5.4 $kcal\ mol^{-1}$ higher than those of the six-membered monocyclic phosphoester, 2-ethoxy-1,3,2-dioxaphosphinane 2-oxide, c.f. Figure 2.48. These calculations supported the hypothesis that the strained 3′,5′-cyclic monomer can promote ring-opening polymerization to afford the resulting poly(3′,5′-cyclic 3-(3-butenyl) thymidine ethyl phosphate)s, c.f. Figure 2.49, with low dispersities smaller than 1.10.

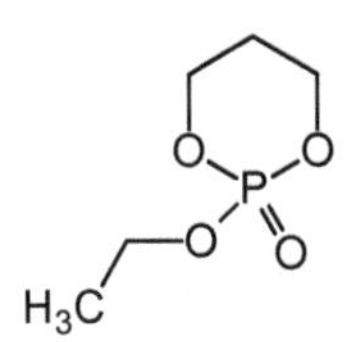

Figure 2.48 2-Ethoxy-1,3,2-dioxaphosphinane 2-oxide.

This design combines the merits of natural product-derived materials and functional, degradable polymers to provide a platform for functional, synthetically derived polydeoxyribonucleotide-analogue materials (200).

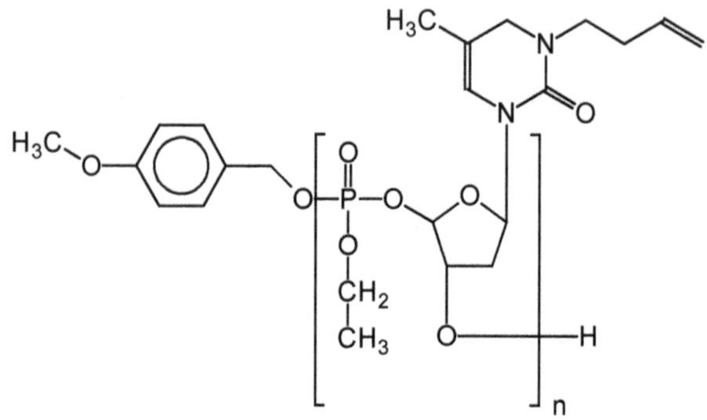

Figure 2.49 Poly(3′,5′-cyclic 3-(3-butenyl) thymidine ethyl phosphate).

2.29.6 *Poly(nucleotide) Compositions*

Functional biomaterials can be used to solve some of the most vexing diagnostic and drug delivery challenges (201). One of the major classes of biomaterials designed to resolve such limitations in treatment is based on functionalized poly(nucleotide)s.

Reversible deactivation radical polymerization (RDRP) methods have been utilized for the preparation of bioconjugates (201). In general, RDRP procedures exhibit tolerance towards functional monomers and functional groups present in nucleic acids and drugs. The three most common RDRP methods are ATRP, nitroxide-mediated polymerization (NMP) and reversible addition-fragmentation transfer RAFT systems, each of which allow unprecedented control over polymer properties such as dimensions (molecular weight), uniformity (polydispersity), topology (geometry), composition and functionality.

There are two methods of conjugating polymers to DNA. In *grafting-to* methods, the DNA and a preformed polymer are conjugated using high-yield linking chemistries, also frequently called *click chemistry*. In *grafting-from* methods, an initiator or transfer agent is immobilized onto DNA and a copolymer is formed through an *in-situ* chain extension polymerization reaction. Technically, polymers conjugated to DNA in this method are graft copolymers but the method has also been referred to as *blocking-from*.

An advantage of *grafting-to* procedures is that the precise composition of the DNA and each polymer segment are known before conjugation. Currently the majority of DNA-polymer conjugates, wherein the polymers are prepared using RDRP, have utilized a *grafting-to* procedure using click chemistries. Even using high-yield click chemistry, however, significant effort must be expended for purification to remove the unreacted reactants.

A method of synthesizing a poly(nucleotide) composition has been presented (201). Here, at least one initiator or one transfer agent is attached for a RDRP to the end of a nucleotide chain assembly. The assembly is immobilized upon a solid-phase support during a solid-phase synthesis of a poly(nucleotide) so that the initiator or the transfer agent is attacked to the end of a nucleotide chain assembly in a manner which is stable under conditions of deprotection of the poly(nucleotide). Also, a polymer grows from the initiator from a site of the chain transfer agent via the RDRP to form the poly(nucleotide) composition.

The poly(nucleotide) can be synthesized via amidite coupling. This method may further include the removal of the poly(nucleotide) from the solid support and deprotection of the poly(nucleotide). The synthesis of a phosphoramidite is shown in Figure 2.50.

The functional phosphoramidite compound can be prepared using commonly available commercial reagents in a two-step procedure, forming a stable amide link between the initiating functionality and the phosphoramidite in good yields (201). This procedure may be employed to incorporate a multiplicity of different functional groups, including groups known to participate in high-yield linking chemistries, into the incorporable phosphoramidite prior to incorporation into nucleic acid molecules at any selected site within the nucleic acid.

The phosphoramidite can be reacted with a free hydroxyl group at the end of the nucleotide chain assembly. The compounds used here are shown in Figure 2.51. The synthesis of phosphoramidite containing an ATRP initiating functionality can be done as follows (201):

Preparation 2–3: First, 5 *g* 2,4-diamino-1-butanol and 6.24 *g* triethylamine were dissolved in 20 *ml* of dichloromethane and 12.8 *g* of α-bromoisobutyryl bromide was added dropwise. The mixture was stirred for 16 *h*. Then, the reaction mixture was filtered and stirred with 20 *ml* of 5% KOH for 2 *h*. The reaction mixture was then added to a separatory

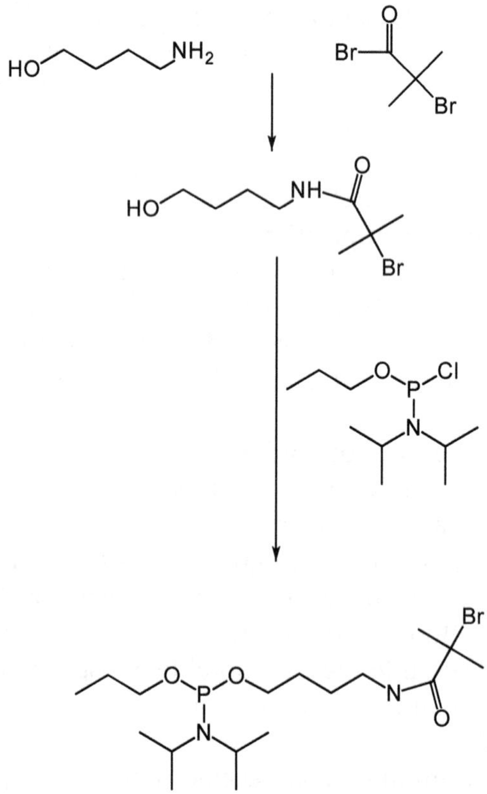

Figure 2.50 Synthesis of a phosphoramidite (201).

funnel and the aqueous layer was separated. The organic layer was then washed with 1N NaOH, 1N HCl, brine dried over $MgSO_4$, filtered and the solvent was evaporated.

Next, 1.24 *ml* *N*,*N*-Diisopropylethylamine, 478 *µl* 2-cyanoethoxy-*N*,*N*-diisopropyl chlorophosphine and 57 *µl* 1-methyl-imidazole were added to a solution of 340 *mg* alcohol in 10 *ml* CH_2Cl_2. The mixture was stirred for 30 *min* at 0°C and 1.5 *h* at room temperature. A workup was done with $NaHCO_3$ (saturated)/ethyl acetate.

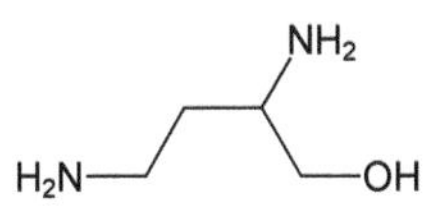

2,4-Diamino-1-butanol

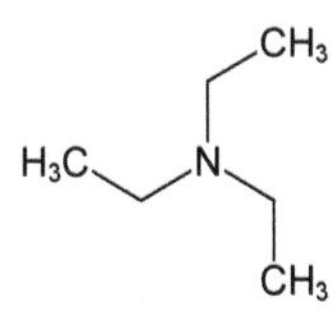

Triethylamine

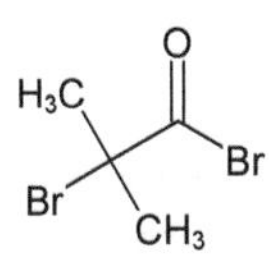

α-Bromoisobutyryl bromide

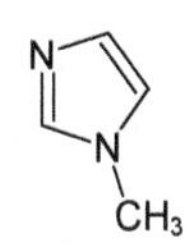

1-Methyl-imidazole

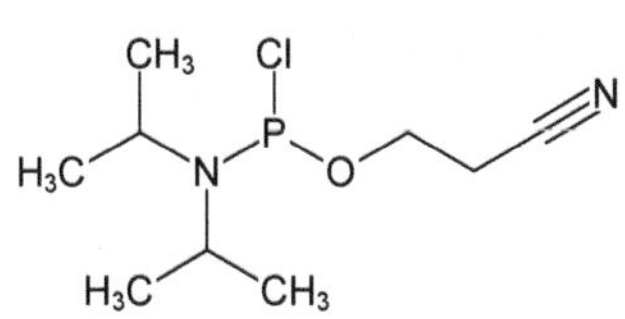

2-Cyanoethoxy-*N*,*N*-diisopropyl chlorophosphine

Figure 2.51 Compounds for the synthesis of phosphoramidite containing an ATRP initiating functionality (201).

In the case of a nitroxide-mediated polymerization, nitroxides suitable as nitroxide-mediated polymerization initiators can be used. Examples are shown in Table 2.6.

Table 2.6 Nitroxide-mediated polymerization initiators (201).

Compound
tert-Butyl-1-phenyl-2-methylpropyl nitroxide
tert-Butyl-1-(2-naphthyl)-2-methylpropyl nitroxide
tert-Butyl-1-diethylphosphono-2,2-dimethylpropyl nitroxide
tert-Butyl-1-dibenzylphosphono-2,2-dimethylpropyl nitroxide
Phenyl-1-diethylphosphono-2,2-dimethylpropyl nitroxide
Phenyl-1-diethylphosphono-1-methylethyl nitroxide
N-*tert*-butyl-*N*-[1-diethylphosphono-(2,2,-dimethylpropyl)]nitroxide
1-Phenyl-2-methylpropyl-1-diethylphosphono-1-methylethylnitroxide

References

1. S.R. Chowdhury and S. Sivaram, *Synthesis of functional polymers of polar and nonpolar monomers by living and/or controlled polymerization* in R. Shunmugam, ed., *Functional Polymers: Design, Synthesis, and Applications*, chapter 1, pp. 3–56. Apple Academic Press, Wareton, NJ, 2017.
2. C. Wesdemiotis, *Angewandte Chemie International Edition*, Vol. 56, p. 1452, 2016.
3. S.D. Hanton, *Chemical Reviews*, Vol. 101, p. 527, 2001.
4. G. Montaudo and R.P. Lattimer, eds., *Mass Spectrometry of Polymers*, CRC Press, Boca Raton, 2002.
5. H. Pasch and W. Schrepp, *MALDI-TOF Mass Spectrometry of Synthetic Polymers*, Springer Berlin Heidelberg, Berlin, Heidelberg, 2003.
6. C. Barner-Kowollik, T. Gruendling, J. Falkenhagen, and S. Weidner, eds., *Mass Spectrometry in Polymer Chemistry*, Wiley-VCH, Weinheim, 2012.
7. V. Gianotti, D. Antonioli, K. Sparnacci, M. Laus, C. Cassino, F. Marsano, G. Seguini, and M. Perego, *Journal of Analytical and Applied Pyrolysis*, Vol. 128, p. 238 , 2017.
8. J. Frechet, *Science*, Vol. 263, p. 1710, 1994.
9. Wikipedia contributors, Stille reaction — Wikipedia, the free encyclopedia, 2018. [Online; accessed 21-August-2018].
10. Z. Bao, W. Chan, and L. Yu, *Chemistry of Materials*, Vol. 5, p. 2, 1993.
11. Z. Bao, W.K. Chan, and L. Yu, *Journal of the American Chemical Society*, Vol. 117, p. 12426, 1995.
12. C. Zhang, T. Matos, R. Li, S.-S. Sun, J.E. Lewis, J. Zhang, and X. Jiang, *Polymer Chemistry*, Vol. 1, p. 663, 2010.
13. T. Zheng, A.M. Schneider, and L. Yu, *Stille polycondensation: A versatile synthetic approach to functional polymers* in *Synthetic Methods for Conjugated Polymers and Carbon Materials*, chapter 1, pp. 1–58. Wiley-Blackwell, 2017.
14. E.H. Discekici, A. Anastasaki, R. Kaminker, J. Willenbacher, N.P. Truong, C. Fleischmann, B. Oschmann, D.J. Lunn, J. Read de Alaniz, T.P. Davis, C.M. Bates, and C.J. Hawker, *Journal of the American Chemical Society*, Vol. 139, p. 5939, 2017.
15. H. Leicht, I. Göttker-Schnetmann, and S. Mecking, *Journal of the American Chemical Society*, Vol. 139, p. 6823, 2017.
16. H.J.H. Fenton, *J. Chem. Soc. Trans.*, Vol. 65, p. 899, 1894.
17. A.E. Shilov and G.B. Shul'pin, *Chemical Reviews*, Vol. 97, p. 2879, 1997.
18. J.A. Labinger, *Chemical Reviews*, Vol. 117, p. 8483, 2017.
19. M.S. Chen and M.C. White, *Science*, Vol. 318, p. 783, 2007.
20. R.H. Crabtree and A. Lei, *Chemical Reviews*, Vol. 117, p. 8481, 2017.
21. C. Ma, P. Fang, and T.-S. Mei, *ACS Catalysis*, Vol. 8, p. 7179, 2018.

22. Y. Wei, P. Hu, M. Zhang, and W. Su, *Chemical Reviews*, Vol. 117, p. 8864, 2017.
23. M.P. Doyle and K.I. Goldberg, *Accounts of Chemical Research*, Vol. 45, p. 777, 2012.
24. H.M.L. Davies and J.R. Manning, *Nature*, Vol. 451, p. 417, January 2008.
25. K.S. Halskov, H.S. Roth, and J.A. Ellman, *Angewandte Chemie*, Vol. 129, p. 9311, 2017.
26. M. Monier, A. El-Mekabaty, and D.A. Abdel-Latif, *Reactive and Functional Polymers*, Vol. 128, p. 104, 2018.
27. J.C. Bohling and K.J. Henderson, Pigmented coating composition with itaconic acid functionalized binder, US Patent 9 464 205, assigned to Rohm and Haas Company (Philadelphia, PA), October 11, 2016.
28. J.K. Bardman and W.T. Brown, Coating with improved hiding, compositions prepared therewith, and processes for the preparation thereof, US Patent 7 081 488, assigned to Rohm and Haas Company (Philadelphia, PA), July 25, 2006.
29. A. Ramos, S. Sousa, D.V. Evtuguin, and J.A. Gamelas, *Reactive and Functional Polymers*, Vol. 117, p. 89, 2017.
30. H. Nosrati, M. Salehiabar, S. Davaran, A. Ramazani, H.K. Manjili, and H. Danafar, *Research on Chemical Intermediates*, Vol. 43, p. 7423, December 2017.
31. J.S. Basuki, A. Jacquemin, L. Esser, Y. Li, C. Boyer, and T.P. Davis, *Polym. Chem.*, Vol. 5, p. 2611, 2014.
32. E. Amstad, T. Gillich, I. Bilecka, M. Textor, and E. Reimhult, *Nano Letters*, Vol. 9, p. 4042, 2009.
33. L. Zhu, D. Wang, X. Wei, X. Zhu, J. Li, C. Tu, Y. Su, J. Wu, B. Zhu, and D. Yan, *Journal of Controlled Release*, Vol. 169, p. 228, 2013. Second Symposium on Innovative Polymers for Controlled Delivery (SIPCD 2012).
34. R. Lakshmanan, M. Sanchez-Dominguez, J.A. Matutes-Aquino, S. Wennmalm, and G. Kuttuva Rajarao, *Langmuir*, Vol. 30, p. 1036, 2014.
35. K. Aurich, M. Schwalbe, J.H. Clement, W. Weitschies, and N. Buske, *Journal of Magnetism and Magnetic Materials*, Vol. 311, p. 15, 2007. Proceedings of the Sixth International Conference on the Scientific and Clinical Applications of Magnetic Carriers.
36. Y.-Y. Liang, L.-M. Zhang, W. Jiang, and W. Li, *ChemPhysChem*, Vol. 8, p. 2367, 2007.
37. P.-C. Lin, P.-H. Chou, S.-H. Chen, H.-K. Liao, K.-Y. Wang, Y.-J. Chen, and C.-C. Lin, *Small*, Vol. 2, p. 485, 2006.

38. J.L. Arias, M. Lopez-Viota, E. Sáez-Fernández, M.A. Ruiz, and Ángel V. Delgado, *Colloids and Surfaces A: Physicochemical and Engineering Aspects*, Vol. 384, p. 157, 2011.
39. C. Hoffmann, V. Chiaula, L. Yu, M. Pinelo, J.M. Woodley, and A.E. Daugaard, *Macromolecular Rapid Communications*, Vol. 39, p. 1700394, 2017.
40. M.M. Nir, A. Ram, and J. Miltz, *Polymer Engineering & Science*, Vol. 35, p. 1878, 1995.
41. A. Mendez-Prieto and S. Sanchez-Valdes, *Journal of Polymer Engineering*, Vol. 18, p. 221, 1998.
42. Y. Tanaka and Y. Hayakawa, Composition of polyamide and acid or anhydride-grafted ethylene/C6–C20 α-olefin copolymer, US Patent 6 075 091, assigned to Mitsui Chemicals Inc. (Tokyo, JP), June 13, 2000.
43. W.J. Harris, J.D. Weaver, B.W. Walther, S.F. Hahn, Y.W. Cheung, P. Gupta, T.H. Ho, K.N. Reichek, S. Yalvac, T.P. Karjala, B.R. Rozenblat, and C.L. Rickey, Functionalized ethylene/α-olefin interpolymer compositions, US Patent 7 897 689, assigned to Dow Global Technologies Inc. (Midland, MI), March 1, 2011.
44. T. Aida, E.W. Meijer, and S.I. Stupp, *Science*, Vol. 335, p. 813, 2012.
45. B. An, X. Wang, M. Cui, X. Gui, X. Mao, Y. Liu, K. Li, C. Chu, J. Pu, S. Ren, Y. Wang, G. Zhong, T.K. Lu, C. Liu, and C. Zhong, *ACS Nano*, Vol. 11, p. 6985, 2017.
46. O.J.G.M. Goor, S.I.S. Hendrikse, P.Y.W. Dankers, and E.W. Meijer, *Chemical Society Reviews*, Vol. 46, p. 6621, 2017.
47. O.J.G.M. Goor, H.M. Keizer, A.L. Bruinen, M.G.J. Schmitz, R.M. Versteegen, H.M. Janssen, R.M.A. Heeren, and P.Y.W. Dankers, *Advanced Materials*, Vol. 29, p. 1604652, 2016.
48. J.D. Hartgerink, E. Beniash, and S.I. Stupp, *Proceedings of the National Academy of Sciences*, Vol. 99, p. 5133, 2002.
49. F.A. Plamper and W. Richtering, *Accounts of Chemical Research*, Vol. 50, p. 131, 2017.
50. M.J.H. Worthington, R.L. Kucera, and J.M. Chalker, *Green Chemistry*, Vol. 19, p. 2748, 2017.
51. H. Kim, J. Lee, H. Ahn, O. Kim, and M.J. Park, *Nature Communications*, Vol. 6, p. 7278, June 2015.
52. S.H. Je, T.H. Hwang, S.N. Talapaneni, O. Buyukcakir, H.J. Kim, J.-S. Yu, S.-G. Woo, M.C. Jang, B.K. Son, A. Coskun, and J.W. Choi, *ACS Energy Letters*, Vol. 1, p. 566, 2016.
53. T.S. Kleine, N.A. Nguyen, L.E. Anderson, S. Namnabat, E.A. LaVilla, S.A. Showghi, P.T. Dirlam, C.B. Arrington, M.S. Manchester, J. Schwiegerling, R.S. Glass, K. Char, R.A. Norwood, M.E. Mackay, and J. Pyun, *ACS Macro Letters*, Vol. 5, p. 1152, 2016.

54. Y. Zhang, J.J. Griebel, P.T. Dirlam, N.A. Nguyen, R.S. Glass, M.E. Mackay, K. Char, and J. Pyun, *Journal of Polymer Science Part A: Polymer Chemistry*, Vol. 55, p. 107, 2016.
55. T.R. Martin, K.A. Mazzio, H.W. Hillhouse, and C.K. Luscombe, *Chem. Commun.*, Vol. 51, p. 11244, 2015.
56. M.K. Salman, B. Karabay, L.C. Karabay, and A. Cihaner, *Journal of Applied Polymer Science*, Vol. 133, 2016.
57. M. Arslan, B. Kiskan, E.C. Cengiz, R. Demir-Cakan, and Y. Yagci, *European Polymer Journal*, Vol. 80, p. 70, 2016.
58. I. Gomez, D. Mecerreyes, J.A. Blazquez, O. Leonet, H.B. Youcef, C. Li, J.L. Gómez-Cámer, O. Bondarchuk, and L. Rodriguez-Martinez, *Journal of Power Sources*, Vol. 329, p. 72, 2016.
59. E.T. Kim, W.J. Chung, J. Lim, P. Johe, R.S. Glass, J. Pyun, and K. Char, *Polym. Chem.*, Vol. 5, p. 3617, 2014.
60. B. Oschmann, J. Park, C. Kim, K. Char, Y.-E. Sung, and R. Zentel, *Chemistry of Materials*, Vol. 27, p. 7011, 2015.
61. P.T. Dirlam, A.G. Simmonds, R.C. Shallcross, K.J. Arrington, W.J. Chung, J.J. Griebel, L.J. Hill, R.S. Glass, K. Char, and J. Pyun, *ACS Macro Letters*, Vol. 4, p. 111, 2015.
62. Z. Sun, M. Xiao, S. Wang, D. Han, S. Song, G. Chen, and Y. Meng, *J. Mater. Chem. A*, Vol. 2, p. 9280, 2014.
63. P.T. Dirlam, A.G. Simmonds, T.S. Kleine, N.A. Nguyen, L.E. Anderson, A.O. Klever, A. Florian, P.J. Costanzo, P. Theato, M.E. Mackay, R.S. Glass, K. Char, and J. Pyun, *RSC Adv.*, Vol. 5, p. 24718, 2015.
64. M.P. Crockett, A.M. Evans, M.J.H. Worthington, I.S. Albuquerque, A.D. Slattery, C.T. Gibson, J.A. Campbell, D.A. Lewis, G.J.L. Bernardes, and J.M. Chalker, *Angewandte Chemie International Edition*, Vol. 55, p. 1714, 2015.
65. I. Gomez, O. Leonet, J.A. Blazquez, and D. Mecerreyes, *ChemSusChem*, Vol. 9, p. 3419, 2016.
66. C. Fu, G. Li, J. Zhang, B. Cornejo, S.S. Piao, K.N. Bozhilov, R.C. Haddon, and J. Guo, *ACS Energy Letters*, Vol. 1, p. 115, 2016.
67. A. Hoefling, Y.J. Lee, and P. Theato, *Macromolecular Chemistry and Physics*, Vol. 218, p. 1600303, 2016.
68. T. Hasell, D.J. Parker, H.A. Jones, T. McAllister, and S.M. Howdle, *Chem. Commun.*, Vol. 52, p. 5383, 2016.
69. D.J. Parker, H.A. Jones, S. Petcher, L. Cervini, J.M. Griffin, R. Akhtar, and T. Hasell, *J. Mater. Chem. A*, Vol. 5, p. 11682, 2017.
70. P. Liu, J.M. Gardner, and L. Kloo, *Chem. Commun.*, Vol. 51, p. 14660, 2015.
71. W.J. Chung, J.J. Griebel, E.T. Kim, H. Yoon, A.G. Simmonds, H.J. Ji, P.T. Dirlam, R.S. Glass, J.J. Wie, N.A. Nguyen, B.W. Guralnick, J. Park, A. Somogyi, P. Theato, M.E. Mackay, Y.-E. Sung, K. Char, and J. Pyun, *Nature Chemistry*, Vol. 5, p. 518, April 2013.

72. P.T. Dirlam, J. Park, A.G. Simmonds, K. Domanik, C.B. Arrington, J.L. Schaefer, V.P. Oleshko, T.S. Kleine, K. Char, R.S. Glass, C.L. Soles, C. Kim, N. Pinna, Y.-E. Sung, and J. Pyun, *ACS Applied Materials & Interfaces*, Vol. 8, p. 13437, 2016.
73. S. Zhuo, Y. Huang, C. Liu, H. Wang, and B. Zhang, *Chem. Commun.*, Vol. 50, p. 11208, 2014.
74. T. Ouchi and A. Fujino, *Die Makromolekulare Chemie*, Vol. 190, p. 1523, 1989.
75. Y. Kimura, K. Shirotani, H. Yamane, and T. Kitao, *Macromolecules*, Vol. 21, p. 3338, 1988.
76. W.W. Gerhardt, D.E. Noga, K.I. Hardcastle, A.J. García, D.M. Collard, and M. Weck, *Biomacromolecules*, Vol. 7, p. 1735, 2006.
77. M. Bednarek, M. Basko, and P. Kubisa, *Reactive and Functional Polymers*, Vol. 119, p. 9, 2017.
78. P. Guerin, M. Vert, C. Braud, and R.W. Lenz, *Polymer Bulletin*, Vol. 14, p. 187, August 1985.
79. B. He, Y. Wan, J. Bei, and S. Wang, *Biomaterials*, Vol. 25, p. 5239, 2004.
80. M. Trollsås, V.Y. Lee, D. Mecerreyes, P. Löwenhielm, M. Möller, R.D. Miller, and J.L. Hedrick, *Macromolecules*, Vol. 33, p. 4619, 2000.
81. I. Fiétier, A. Le Borgne, and N. Spassky, *Polymer Bulletin*, Vol. 24, p. 349, October 1990.
82. X.-Q. Liu, Z.-C. Li, F.-S. Du, and F.-M. Li, *Macromolecular Rapid Communications*, Vol. 20, p. 470, 1999.
83. X.-Q. Liu, M.-X. Wang, Z.-C. Li, and F.-M. Li, *Macromolecular Chemistry and Physics*, Vol. 200, p. 468, 1999.
84. D. Tian, P. Dubois, C. Grandfils, and R. Jérôme, *Macromolecules*, Vol. 30, p. 406, 1997.
85. D. Tian, P. Dubois, and R. Jérôme, *Macromolecules*, Vol. 30, p. 2575, 1997.
86. D. Tian, P. Dubois, and R. Jérôme, *Macromolecules*, Vol. 30, p. 1947, 1997.
87. D. Tian, O. Halleux, P. Dubois, R. Jérôme, R. Sobry, and G. Van den Bossche, *Macromolecules*, Vol. 31, p. 924, 1998.
88. F. Stassin, O. Halleux, P. Dubois, C. Detrembleur, P. Lecomte, and R. Jérôme, *Macromolecular Symposia*, Vol. 153, p. 27, 2000.
89. D. Mecerreyes, B. Atthoff, K.A. Boduch, M. Trollsås, and J.L. Hedrick, *Macromolecules*, Vol. 32, p. 5175, 1999.
90. P.J.A. In't Veld, P.J. Dijkstra, and J. Feijen, *Die Makromolekulare Chemie*, Vol. 193, p. 2713, 1992.
91. D.J. Boday and T.C. Mauldin, Lactide-functionalized polymer, US Patent 9 458 268, assigned to International Business Machines Corporation (Armonk, NY), October 4, 2016.
92. S. Sun and S. Liang, *Nanoscale*, Vol. 9, p. 10544, 2017.

93. M. Groenewolt and M. Antonietti, *Advanced Materials*, Vol. 17, p. 1789, 2005.
94. F. Goettmann, A. Fischer, M. Antonietti, and A. Thomas, *Angewandte Chemie International Edition*, Vol. 45, p. 4467, 2006.
95. Y. Wang, X. Wang, and M. Antonietti, *Angewandte Chemie International Edition*, Vol. 51, p. 68, 2011.
96. S. Cao, J. Low, J. Yu, and M. Jaroniec, *Advanced Materials*, Vol. 27, p. 2150, 2015.
97. Y. Zheng, L. Lin, B. Wang, and X. Wang, *Angewandte Chemie International Edition*, Vol. 54, p. 12868, 2015.
98. Y. Gong, M. Li, H. Li, and Y. Wang, *Green Chem.*, Vol. 17, p. 715, 2015.
99. S. Yin, J. Han, T. Zhou, and R. Xu, *Catalysis Science & Technology*, Vol. 5, p. 5048, 2015.
100. Z. Zhao, Y. Sun, and F. Dong, *Nanoscale*, Vol. 7, p. 15, 2015.
101. Q. Han, B. Wang, J. Gao, Z. Cheng, Y. Zhao, Z. Zhang, and L. Qu, *ACS Nano*, Vol. 10, p. 2745, 2016.
102. L. Jiang, X. Yuan, Y. Pan, J. Liang, G. Zeng, Z. Wu, and H. Wang, *Applied Catalysis B: Environmental*, Vol. 217, p. 388, 2017.
103. R.R. Haikal, A.B. Soliman, P.J. Pellechia, S. Heißler, M. Tsotsalas, S.S. Ali, and M.H. Alkordi, *Carbon*, Vol. 118, p. 215, 2017.
104. T. Xu, K. Tu, J. Cheng, Y. Ni, L. Zhang, Z. Cheng, and X. Zhu, *Macromolecular Rapid Communications*, Vol. 39, p. 1800151, 2018.
105. W. Zhang, G. Zhang, L. Du, C. Zhang, L. Li, J. Zhu, J. Pei, and J. Wu, *Reactive and Functional Polymers*, Vol. 127, p. 161, 2018.
106. H. Tian, Z. Tang, X. Zhuang, X. Chen, and X. Jing, *Progress in Polymer Science*, Vol. 37, p. 237, 2012. 37/2 Topical Issue on Biomaterials.
107. K.J. Zhu, R.W. Hendren, K. Jensen, and C.G. Pitt, *Macromolecules*, Vol. 24, p. 1736, 1991.
108. G. Rokicki, *Progress in Polymer Science*, Vol. 25, p. 259, 2000.
109. X. Chen, S.P. McCarthy, and R.A. Gross, *Macromolecules*, Vol. 30, p. 3470, 1997.
110. P. Zinck, F. Bonnet, A. Mortreux, and M. Visseaux, *Progress in Polymer Science*, Vol. 34, p. 369, 2009.
111. S. Harder, F. Feil, and K. Knoll, *Angewandte Chemie International Edition*, Vol. 40, p. 4261, 2001.
112. J. Li and H.-M. Li, *European Polymer Journal*, Vol. 41, p. 823, 2005.
113. Y. Gao and H.-M. Li, *Polymer International*, Vol. 53, p. 1436, 2005.
114. Y. Gao, H.-M. Li, F.-S. Liu, X.-Y. Wang, and Z.-G. Shen, *Journal of Polymer Research*, Vol. 14, p. 291, August 2007.
115. E.B. Orler, D.J. Yontz, and R.B. Moore, *Macromolecules*, Vol. 26, p. 5157, 1993.
116. E.B. Orler and R.B. Moore, *Macromolecules*, Vol. 27, p. 4774, 1994.
117. H.-M. Li, J.-C. Liu, F.-M. Zhu, and S.-A. Lin, *Polymer International*, Vol. 50, p. 421, 2001.

118. J. Shin, S.M. Jensen, J. Ju, S. Lee, Z. Xue, S.K. Noh, and C. Bae, *Macromolecules*, Vol. 40, p. 8600, 2007.
119. P. Leophairatana, S. Samanta, C.C. De Silva, and J.T. Koberstein, *Journal of the American Chemical Society*, Vol. 139, p. 3756, 2017.
120. Wikipedia contributors, Glaser coupling — Wikipedia, the free encyclopedia, https://en.wikipedia.org/w/index.php?title=Glaser_coupling&oldid=840645568, 2018. [Online; accessed 23-August-2018].
121. H. Uyama, *Polymer Journal*, p. 1, 2018.
122. T. Philibert, B.H. Lee, and N. Fabien, *Applied Biochemistry and Biotechnology*, Vol. 181, p. 1314, April 2017.
123. T. Nishimura and K. Akiyoshi, *Wiley Interdisciplinary Reviews: Nanomedicine and Nanobiotechnology*, Vol. 9, p. e1423, 2016.
124. G. Ziegast and B. Pfannemüller, *Carbohydrate Research*, Vol. 160, p. 185, 1987.
125. T. Tanaka, S. Sasayama, S. Nomura, K. Yamamoto, Y. Kimura, and J.-I. Kadokawa, *Macromolecular Chemistry and Physics*, Vol. 214, p. 2829, 2013.
126. R. Rachmawati, A.J.J. Woortman, K. Kumar, and K. Loos, *Macromolecular Bioscience*, Vol. 15, p. 812, 2015.
127. Y. Yataka, T. Sawada, and T. Serizawa, *Chemical Communications*, Vol. 51, p. 12525, 2015.
128. S. Toita, N. Morimoto, and K. Akiyoshi, *Biomacromolecules*, Vol. 11, p. 397, 2010.
129. S. Toita, Y. Soma, N. Morimoto, and K. Akiyoshi, *Chemistry Letters*, Vol. 38, p. 1114, 2009.
130. Y. Tahara, J. Yasuoka, S. Sawada, Y. Sasaki, and K. Akiyoshi, *Biomater. Sci.*, Vol. 3, p. 256, 2015.
131. H. Fujii, M. Shin-Ya, S. Takeda, Y. Hashimoto, S.-A. Mukai, S.-I. Sawada, T. Adachi, K. Akiyoshi, T. Miki, and O. Mazda, *Cancer Science*, Vol. 105, p. 1616, 2014.
132. S. Toita, S.-I. Sawada, and K. Akiyoshi, *Journal of Controlled Release*, Vol. 155, p. 54 , 2011.
133. D.N. Crisan, O. Creese, R. Ball, J.L. Brioso, B. Martyn, J. Montenegro, and F. Fernandez-Trillo, *Polymer Chemistry*, Vol. 8, p. 4576, 2017.
134. A.B. Lowe and C.L. McCormick, *Progress in Polymer Science*, Vol. 32, p. 283, 2007.
135. C. Barner-Kowollik, ed., *Handbook of RAFT Polymerization*, Wiley-VCH, Weinheim, 2008.
136. C. Boyer, V. Bulmus, T.P. Davis, V. Ladmiral, J. Liu, and S. Perrier, *Chemical Reviews*, Vol. 109, p. 5402, 2009.
137. J.M. Priegue, D.N. Crisan, J. Martínez-Costas, J.R. Granja, F. Fernandez-Trillo, and J. Montenegro, *Angewandte Chemie International Edition*, Vol. 55, p. 7492, 2016.

138. H.-Y. Wang, L. Cui, J.-Z. Xie, C.F. Leong, D.M. D'Alessandro, and J.-L. Zuo, *Coordination Chemistry Reviews*, Vol. 345, p. 342, 2017.
139. S.R. Batten, N.R. Champness, X.-M. Chen, J. Garcia-Martinez, S. Kitagawa, L. Öhrström, M. O'Keeffe, M.P. Suh, and J. Reedijk, *CrystEngComm*, Vol. 14, p. 3001, 2012.
140. S.R. Batten, N.R. Champness, X.-M. Chen, J. Garcia-Martinez, S. Kitagawa, L. Öhrström, M. O'Keeffe, M.P. Suh, and J. Reedijk, *Pure and Applied Chemistry*, Vol. 85, p. 1715, 2013.
141. J.R. Long and O.M. Yaghi, *Chem. Soc. Rev.*, Vol. 38, p. 1213, 2009.
142. H.-C. Zhou, J.R. Long, and O.M. Yaghi, *Chemical Reviews*, Vol. 112, p. 673, 2012.
143. H.-C.J. Zhou and S. Kitagawa, *Chem. Soc. Rev.*, Vol. 43, p. 5415, 2014.
144. Y. Mitamura, H. Yorimitsu, K. Oshima, and A. Osuka, *Chem. Sci.*, Vol. 2, p. 2017, 2011.
145. Y. Ji, L. Zhang, X. Gu, W. Zhang, N. Zhou, Z. Zhang, and X. Zhu, *Angewandte Chemie International Edition*, Vol. 56, p. 2328, 2017.
146. L. Yu, Z. Zhang, Y.-Z. You, and C.-Y. Hong, *European Polymer Journal*, Vol. 103, p. 80, 2018.
147. Y. Li, X.-H. Dong, Y. Zou, Z. Wang, K. Yue, M. Huang, H. Liu, X. Feng, Z. Lin, W. Zhang, W.-B. Zhang, and S.Z. Cheng, *Polymer*, Vol. 125, p. 303, 2017.
148. M. Paeth, J. Stapleton, M.L. Dougherty, H. Fischesser, J. Shepherd, M. McCauley, R. Falatach, R.C. Page, J.A. Berberich, and D. Konkolewicz, *Approaches for conjugating tailor-made polymers to proteins* in C.V. Kumar, ed., *NanoArmoring of Enzymes: Rational Design of Polymer-Wrapped Enzymes*, Vol. 590 of *Methods in Enzymology*, chapter 9, pp. 193–224. Academic Press, 2017.
149. J.-F. Lutz, M. Ouchi, D.R. Liu, and M. Sawamoto, *Science*, Vol. 341, 2013.
150. N. ten Brummelhuis, *Polymer Chemistry*, Vol. 6, p. 654, 2015.
151. J.-F. Lutz, J.-M. Lehn, E.W. Meijer, and K. Matyjaszewski, *Nature Reviews Materials*, Vol. 1, p. 16024, April 2016.
152. N. ten Brummelhuis, P. Wilke, and H.G. Börner, *Macromolecular Rapid Communications*, Vol. 38, p. 1700632, 2017.
153. S.C. Solleder, D. Zengel, K.S. Wetzel, and M.A.R. Meier, *Angewandte Chemie International Edition*, Vol. 55, p. 1204, 2015.
154. Wikipedia contributors, Passerini reaction — Wikipedia, the free encyclopedia, https://en.wikipedia.org/w/index.php?title=Passerini_reaction&oldid=787461936, 2017. [Online; accessed 21-August-2018].
155. M. Passerini, *Gazz. Chim. Ital.*, Vol. 51, p. 181, 1921.
156. Y. Hibi, M. Ouchi, and M. Sawamoto, *Nature Communications*, Vol. 7, p. 11064, March 2016.

157. J.P. Cole, A.M. Hanlon, K.J. Rodriguez, and E.B. Berda, *Journal of Polymer Science Part A: Polymer Chemistry*, Vol. 55, p. 191, 2016.
158. R.M. Weiss, A.L. Short, and T.Y. Meyer, *ACS Macro Letters*, Vol. 4, p. 1039, 2015.
159. B.V.K.J. Schmidt, N. Fechler, J. Falkenhagen, and F. Lutz, *Nature Chemistry*, Vol. 3, p. 234, January 2011.
160. R.K. Roy and J.-F. Lutz, *Journal of the American Chemical Society*, Vol. 136, p. 12888, 2014.
161. C.K. Lyon, A. Prasher, A.M. Hanlon, B.T. Tuten, C.A. Tooley, P.G. Frank, and E.B. Berda, *Polym. Chem.*, Vol. 6, p. 181, 2015.
162. M. Huo, N. Wang, T. Fang, M. Sun, Y. Wei, and J. Yuan, *Polymer*, Vol. 66, p. A11 , 2015.
163. A.M. Hanlon, C.K. Lyon, and E.B. Berda, *Macromolecules*, Vol. 49, p. 2, 2016.
164. H. Kitagishi, K. Oohora, H. Yamaguchi, H. Sato, T. Matsuo, A. Harada, and T. Hayashi, *Journal of the American Chemical Society*, Vol. 129, p. 10326, 2007.
165. M. Raynal, P. Ballester, A. Vidal-Ferran, and P.W.N.M. van Leeuwen, *Chem. Soc. Rev.*, Vol. 43, p. 1734, 2014.
166. E. Huerta, B. van Genabeek, B.A.G. Lamers, M.M.E. Koenigs, E.W. Meijer, and A.R.A. Palmans, *Chemistry - A European Journal*, Vol. 21, p. 3682, 2015.
167. C.A. Tooley, S. Pazicni, and E.B. Berda, *Polym. Chem.*, Vol. 6, p. 7646, 2015.
168. K.J. Rodriguez, A.M. Hanlon, C.K. Lyon, J.P. Cole, B.T. Tuten, C.A. Tooley, E.B. Berda, and S. Pazicni, *Inorganic Chemistry*, Vol. 55, p. 9493, 2016.
169. C.R.D. Lancaster, *FEBS Letters*, Vol. 545, p. 52, 2003.
170. T.L. Poulos, *Chemical Reviews*, Vol. 114, p. 3919, 2014.
171. J. Sun, X. Jiang, A. Siegmund, M.D. Connolly, K.H. Downing, N.P. Balsara, and R.N. Zuckermann, *Macromolecules*, Vol. 49, p. 3083, 2016.
172. H. Kawaguchi, *Progress in Polymer Science*, Vol. 25, p. 1171, 2000.
173. H. Shen, X. Du, X. Ren, Y. Xie, X. Sheng, and X. Zhang, *Reactive and Functional Polymers*, Vol. 112, p. 53, 2017.
174. J.A. Reglero Ruiz, M. Trigo-López, F.C. García, and J.M. García, *Polymers*, Vol. 9, 2017.
175. K. Matsumoto, K. Nishi, K. Ando, and M. Jikei, *Polym. Chem.*, Vol. 6, p. 4758, 2015.
176. H. Tsushima, K. Matsumoto, and M. Jikei, *Polymers for Advanced Technologies*, Vol. 22, p. 1292, 2011.
177. S. Shabbir, S. Zulfiqar, and M.I. Sarwar, *Journal of Polymer Research*, Vol. 18, p. 1919, November 2011.
178. X. Hu, *Journal of Applied Polymer Science*, Vol. 134, 2016.

179. S.Y. Park, S.G. Kim, J.H. Chun, B.-H. Chun, and S.H. Kim, *Desalination and Water Treatment*, Vol. 43, p. 221, 2012.
180. N.W. Ding, W. Lin, W.L. Sun, and Z.Q. Shen, *Science China Chemistry*, Vol. 54, p. 320, February 2011.
181. G.-S. Liou, H.-Y. Lin, and H.-J. Yen, *J. Mater. Chem.*, Vol. 19, p. 7666, 2009.
182. J. Yu, X. Huang, L. Wang, P. Peng, C. Wu, X. Wu, and P. Jiang, *Polym. Chem.*, Vol. 2, p. 1380, 2011.
183. R. Qian, J. Yu, L. Xie, Y. Li, and P. Jiang, *Polymers for Advanced Technologies*, Vol. 24, p. 348, 2013.
184. K.G. Nilsson, *Journal of Immunological Methods*, Vol. 122, p. 273, 1989.
185. M. Yasui, T. Shiroya, K. Fujimoto, and H. Kawaguchi, *Colloids and Surfaces B: Biointerfaces*, Vol. 8, p. 311, 1997.
186. T. Ikai, *Polymer Journal*, Vol. 49, p. 355, January 2017.
187. H.U. Blaser, *Chemical Reviews*, Vol. 92, p. 935, 1992.
188. D.J. Ager, I. Prakash, and D.R. Schaad, *Chemical Reviews*, Vol. 96, p. 835, 1996.
189. D. Seebach, A.K. Beck, and A. Heckel, *Angewandte Chemie International Edition*, Vol. 40, p. 92, 2001.
190. C.W.Y. Chung and P.H. Toy, *Tetrahedron: Asymmetry*, Vol. 15, p. 387, 2004.
191. H. Pellissier, *Tetrahedron*, Vol. 8, p. 1619, 2006.
192. S.M. Lait, D.A. Rankic, and B.A. Keay, *Chemical Reviews*, Vol. 107, p. 767, 2007.
193. L.M. Geary and P.G. Hultin, *Tetrahedron: Asymmetry*, Vol. 20, p. 131, 2009.
194. M.M. Heravi, V. Zadsirjan, and B. Farajpour, *RSC Adv.*, Vol. 6, p. 30498, 2016.
195. T. Ikai, K. Kimura, K. Maeda, and S. Kanoh, *Reactive and Functional Polymers*, Vol. 82, p. 52, 2014.
196. A. Suzuki, *Chem. Commun.*, pp. 4759–4763, 2005.
197. A.Z. Yu, J.M. Sahouani, R.A. Setien, and D.C. Webster, *Reactive and Functional Polymers*, Vol. 128, p. 29, 2018.
198. X. Zhang, Q. Huang, F. Deng, H. Huang, Q. Wan, M. Liu, and Y. Wei, *Applied Materials Today*, Vol. 7, p. 222 , 2017.
199. V.M. Chernyshev, O.A. Kravchenko, and V.P. Ananikov, *Russian Chemical Reviews*, Vol. 86, p. 357, 2017.
200. Y.-Y.T. Tsao and K.L. Wooley, *Journal of the American Chemical Society*, Vol. 139, p. 5467, 2017.
201. S.R. Das, S. Averick, S.K. Dey, and K. Matyjaszewski, Functionalized polymer hybrids, US Patent 9 765 169, assigned to Carnegie Mellon University (Pittsburgh, PA), September 19, 2017.

3

Technical Applications

3.1 Electrical Applications

3.1.1 *Supercapacitors*

Organic conjugated polymer-based nanohybrids with inorganic metallic nanoparticles and/or nanostructured carbons, such as, graphene, carbon nanotubes, porous carbons, and carbon quantum dots, have been found to exhibit intriguing multifunctional properties for various potential applications in energy storage, e.g., supercapacitors, batteries, energy conversion, solar cells, and electrochemical sensors (1).

The synthesis methodologies of nanostructures of conjugated polymers have been discussed along with their hybrid composites, and the formation mechanism of such nanoscale structures. Furthermore, potential applications as electrochemical supercapacitors and as solar energy materials have been detailed (1).

3.1.1.1 *Carbon-Based Electrodes*

Carbon-based electrode materials with high nitrogen content showed high volumetric specific capacitance and high energy density (2).

Nitrogen-doped porous carbon/multiwalled carbon nanotubes (CNTs) with high nitrogen content could be prepared by directly pyrolyzing a metal-organic coordination polymer composite, which was prepared by the reaction of 4,4′-bipyridine, $FeCl_3$ and multiwalled CNTs. The so-prepared materials exhibit a hierarchical pore structure with a high specific surface area. The assembled symmetric supercapacitors have a superior energy density of 18.82 $Wh\,kg^{-1}$.

3.1.2 *Solar Energy Conversion*

Low energy gap and fully regioregular conjugated polymers find wide use in solar energy conversion applications. Such types of polymers have been reviewed and also the synthesis and characterization of a specific example of a new polymer has been reported (3).

This is a low energy gap, fully regioregular, terminal functionalized, and processable conjugated polymer, i.e., poly(3-dodecyloxy-2,5-thienylene vinylene). The polymer exhibited an optical energy gap of 1.46 eV based on the UV-Vis-NIR absorption spectrum. The electrochemically measured highest occupied molecular orbital (HOMO) level is $-4.79\ eV$, resulting in the lowest unoccupied molecular orbital (LUMO) level of $-3.33\ eV$ based on optical energy gap.

This polymer could be synthesized via Horner-Emmons condensation and is fairly soluble in common organic solvents such as tetrahydrofuran (THF) and chloroform with gentle heating. DSC showed two endothermic peaks at 67°C and 227°C that can be attributed to transitions between crystalline and liquid states. The polymer is thermally stable up to about 300°C. This polymer appears to be very promising for cost-effective solar cell applications (3).

3-Methoxythiophene, c.f. Figure 3.1, was synthesized from 3-bromothiophene and sodium methoxide in the presence of copper bromide. Then, 3-dodecyloxythiophene was synthesized as follows (3):

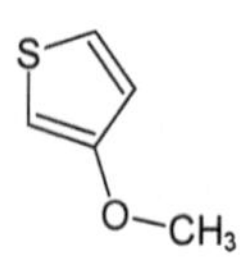

Figure 3.1 3-Methoxythiophene.

Preparation 3–1: First, 235 *mmol* 3-methoxythiophene, 4.78 *mmol* *p*-toluenesulfonic acid monohydrate, c.f. Figure 3.2, and 210.4 *mmol* 1-dodecanol were mixed together in a 1000 *ml* round-bottom flask and refluxed at 92°C at low pressure for 3.5 *h*. The reaction was allowed to cool down to room temperature, followed by the addition of a 1:1 molar equivalent of sodium carbonate to *mmol* toluenesulfonic acid in order to quench the acid. The solution was then washed with 100 *ml* water and 100 *ml* hexane. Subsequent washings of the aqueous and organic layer were done with hexane and water. The separation of the organic layer was done by low pressure to vacuum distillation. A final purification was performed by column chromatography with hexane as eluent. The product was a light green solid with a yield of 91.64%.

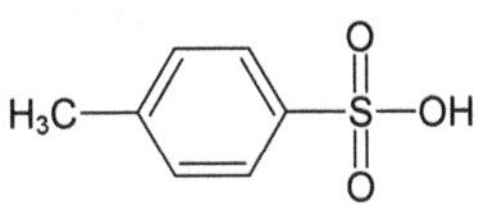

Figure 3.2 *p*-Toluenesulfonic acid.

Also, the syntheses of related compounds have been detailed (3).

3.1.2.1 *Panchromatic Ternary Photovoltaic Cells*

There have been significant recent advances in organic solar cells in the development of nonfullerene acceptors.

The potential advantages of these materials over fullerenes include easier and lower cost synthesis and purification, stronger absorption, and easier tunability of absorption bands and electron affinities via structural changes.

Indacenodithiophene-based nonfullerene acceptors have attracted particular interest, owing to their high electron mobilities and strong absorption at ca. 600 *nm* to 800 *nm*. The basic structure of indacenodithiophene is shown in Figure 3.3.

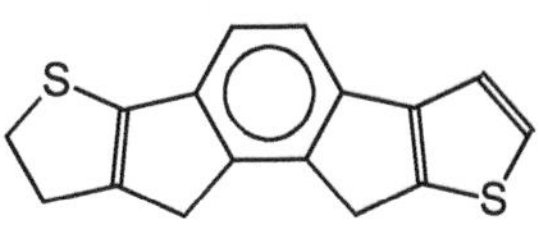

Figure 3.3 Indacenodithiophene.

Panchromatic ternary photovoltaic cells have been synthesized using a nonfullerene acceptor synthesized by C–H functionalization (4).

An economical approach to the synthesis uses a highly regioselective double C–H functionalization of 2,6-naphthalic acid. Carboxylate-directed bromination conditions resulted in 3,7-dibromo-2,6-naphthalic acid, which was then esterified, and coupled with 2-(tributylstannyl)thiophene, c.f. Figure 3.4, to give the desired compound in 62% overall yield. Bromination adjacent to the carboxylic acid is attributable to chelation forming five-membered palladacycle intermediates, whereas 1,5-dibromination is presumably precluded by steric interactions of the Pd with the 4,8-hydrogen atoms (4). The synthesis is shown in Figure 3.5.

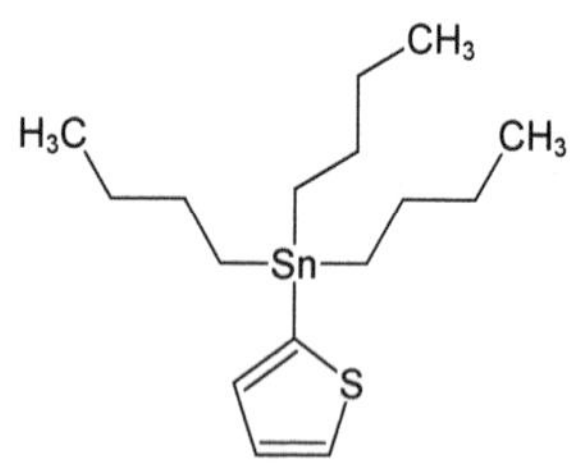

Figure 3.4 2-(Tributylstannyl)thiophene.

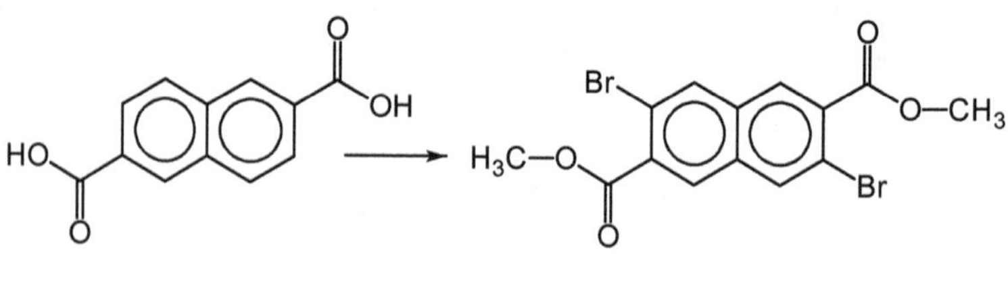

Figure 3.5 Synthesis of a carboxylic acid-directed 3,7-dibromination and cross-coupling (4).

3.1.3 *Light-Emitting Device*

Various nanopatterning methods have been developed. Among these, the two most promising methods at present are extreme ultraviolet lithography and electron beam lithography.

Extreme ultraviolet lithography is an imaging method using very short wavelength ultraviolet light of, e.g., 11 *nm* to 14 *nm*, which may be used to form a fine patterned surface relief grating on a substrate. Extreme ultraviolet lithography uses conceptually similar equipment for effecting the imaging as that used in conventional light lithographic methods.

However, the mask and lens used in conventional light lithographic methods are not useful in extreme ultraviolet lithography due to the smaller feature sizes, which require an improved mask technology, and the high absorbance of conventional lens materials at extreme ultraviolet wavelengths.

Therefore, in place of a lens, extreme ultraviolet lithography uses reflective optics based on higher-order reflective materials, such as, for example, molybdenum silicide layered reflective lenses, and consequently the optics used in extreme ultraviolet lithography are not trivial in their construction or necessarily small-sized.

Electron beam lithography is another method in which a pattern is formed by irradiating an electron beam directly on a substrate. A finer pattern can be formed by electron beam lithography in comparison with that obtained using extreme ultraviolet lithography. However, since electron beam lithography has increased the manufacturing costs and it is difficult to achieve accurate results with it, electron beam lithography is not commonly used in a method for manufacturing an element.

In addition to the methods described above, other non-optical or irradiative methods have been examined, such as, for example, a nanoimprint method in which a pattern is formed using a nano-size mold, a self-assembly method using self-assembling properties of the molecules or a molecular or atomic manipulation in which a pattern is formed by directly manipulating a molecule or an atom. However, for these methods of forming patterns, the materials used are limited and the reproducibility is poor. In particular, it is difficult to form a pattern having a pattern dimension of 1 μm or less.

A method has been described for forming a fine pattern with a dimension of 1 μm or less (5). The method for forming the fine pattern consists of the following steps:

1. Forming an azobenzene-functionalized polymer film on an etched layer,
2. Irradiating the azobenzene-functionalized polymer film using an interference laser beam to form a patterned azobenzene-functionalized polymer film having fine-patterned surface relief gratings by a photophysical mass transporting of the azobenzene-functionalized polymer,
3. Etching the etched layer using the azobenzene-functionalized polymer film having the surface relief grating patterns as an etching mask, and
4. Removing the patterned azobenzene-functionalized polymer film.

Figure 3.6 is a schematic view illustrating an exemplary photoisomerization reaction of an azobenzene-functionalized polymer.

An azobenzene functionality is an isomerizable structure in which the two benzene functionalities are connected by two doubly bonded nitrogen atoms, and is an aromatic compound in which electrons are delocalized, i.e., they are distributed across an entire molecule by a side-overlapping of *p*-orbitals.

An azobenzene-functionalized polymer has a perpendicular orientation to incident light, providing a polarization property for the azobenzene-functionalized polymer.

When the azobenzene-functionalized polymer with an initial random orientation of the azobenzene functional groups is exposed to linear-polarized light, it isomerizes by inversion and is oriented perpendicular to the direction of the linear polarization of incident light.

Based on the perpendicular orientation to incident light polarization property, when an interference laser beam is irradiated onto a surface of the azobenzene-functionalized polymer film, a photophysical mass transporting of the azobenzene-functionalized polymer attributable to the change in dimension and volume of the azobenzene functional groups, which in turn causes a change in volume in the irradiated areas, can be induced thereby, and thus form a relief pattern in the exposed areas. As a result, the surface

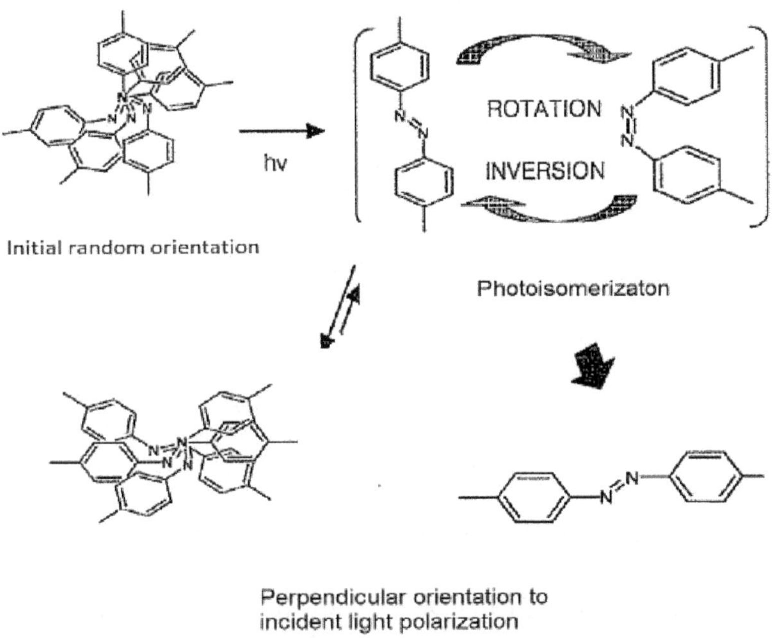

Figure 3.6 Photoisomerization reaction of an azobenzene-functionalized polymer (5).

relief gratings having a fine pattern can be formed at the surface of the azobenzene-functionalized polymer.

A copolymer containing poly disperse orange 3 (PDO3), as shown in Figure 3.7, was employed as the azobenzene-functionalized polymer.

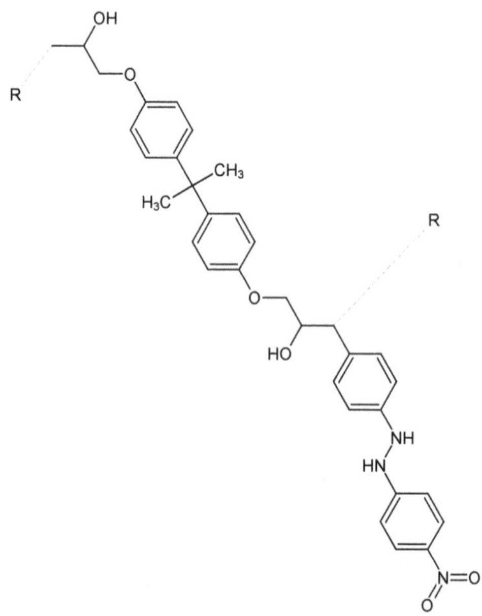

Figure 3.7 Copolymer with poly disperse orange (5).

A solution, in which PDO3 was dissolved in cyclohexanone, was spin-coated on a substrate having an ITO layer deposited on a surface thereof to form a resulting layer having a thickness of 500 μm. The resulting layer was dried in a vacuum oven to remove the cyclohexanone organic solvent from the resulting layer at a temperature of 100°C to form a PDO3 film. The PDO3 film was exposed to an Ar laser at a wavelength of 488 *nm* for 1 *h* to form the surface relief gratings on the PDO3 film.

Next, the ITO layer was etched using the patterned PDO3 film of the surface relief gratings as the etching mask. To etch the ITO layer, an ICP etching method was used, and was performed at a power output of 1000 *W* for 2 *min*. In the ICP etching method, Ar gas and methane gas in a volume ratio of 9:1 were mixed and used as the source gas. The resulting layer was heat-treated at a temperature between ambient temperature and 425°C to remove the patterned azobenzene-functionalized polymer film. In particular, the heat treatment included: Raising the temperature from ambient temperature to 100°C over 30 *min* and maintaining for 1 *h*, raising the temperature from 100°C to 350°C over 30 *min* and maintaining for 5 *h*, and raising the temperature from 300°C to 425°C over 30 *min* and maintaining for 2 *h* (5).

Similar polymers were used before for hybrid solar cells (6). Also, the synthesis has been detailed (7).

3.1.4 *Triboelectric Nanogenerator*

Cellulose, the most abundant natural polymer, is renewable, biodegradable, and cost competitive. The development of a high-performance triboelectric nanogenerator with both contacting materials made from cellulosic materials has been reported (8).

Cellulose nanofibrils are used as the raw material, and chemical reaction approaches were employed to attach nitro groups and methyl groups to cellulose molecules to change the tribopolarities of cellulose nanofibril, which in turn significantly enhances the triboelectric output.

Specifically, the nitrocellulose nanofibril possesses a negative surface charge density of 85.8 $\mu C\, m^{-1}$, while the methyl-cellulose nanofibril possesses a positive surface charge density of 62.5 $\mu C\, m^{-1}$, reaching 71% and 52% of that for fluorinated ethylene propylene, respectively. The figure of merit of the nitrocellulose nanofibril and methyl-cellulose nanofibril is quantified to be 0.504 and 0.267, respectively, comparable to or exceeding a number of common synthetic polymers such as Kapton, poly(vinylidene fluoride), and poly(ethylene). The triboelectric nanogenerator fabricated from nitrocellulose nanofibril paired with methyl-cellulose nanofibril demonstrates an average voltage output of 8 *V* and current output of 9 μA, which approaches the same level obtained from a triboelectric nanogenerator made from fluorinated ethylene propylene.

In summary, a successful strategy of using environmentally friendly, abundant cellulosic materials for replacing the synthetic polymers in the development of a triboelectric nanogenerator could be demonstrated (8).

3.1.5 Conductive Photoresist

A methacryl ethyl-functionalized soluble poly(pyrrole) (PPY) was designed and prepared by chemical oxidative polymerization for rapid fabrication of high-aspect-ratio pillar arrays (9).

The chemical structures of the pyrrole derivatives and the corresponding PPY were characterized by Fourier transform infrared (FTIR) and 1H nuclear magnetic resonance spectroscopy (NMR). The PPY with a weight-average molecular weight of 7.38 *k Dalton* exhibits a good solubility in several organic solvents, favorable thin film-forming ability and two UV-Vis absorption peaks at 280 *nm* and 380 *nm* in THF solution. The dilute chloroform solution of the PPY is a Newtonian fluid with a low viscosity and shows a significant increase in the electrical conductivity with increased content of PPY.

In addition, an insulating photoresist can be transformed into a conductive photoresist by doping the PPY. Electrowetting-driven structure formation experiments could confirm that the conductive photoresist can fulfill rapid fabrication of higher-aspect-ratio pillar arrays in comparison to the insulating photoresist (9).

3.2 Photocatalytic Methods

3.2.1 Oxygen Evolution

Conjugated polymers are emerging as appealing light harvesters for photocatalytic water splitting owing to their adjustable band gap and facile processing (10). An advanced mild synthesis of three conjugated triazine-based polymers with different chain lengths by increasing the quantity of electron-donating benzyl units in the backbone has been reported.

Varying the chain length of the triazine-based polymers modulates their electronic, optical, and redox properties, resulting in an enhanced performance for photocatalytic oxygen evolution, which is the more challenging half-reaction of water splitting owing to the sluggish reaction kinetics (10).

3.2.2 *Water Splitting*

The conversion of solar energy into storable and transportable chemical fuels using artificial photosynthetic systems can provide an alternative route to the current unsustainable use of fossil fuels, addressing the worldwide energy crisis and environmental issues (11).

Semiconducting polymers have emerged as a very promising class of photocatalysts for water splitting as their electronic and structural properties can be conveniently controlled and systematically designed at a molecular level.

Among the various polymer photocatalysts that have been already reported, 2D polymer nanosheets are particularly interesting and are gaining much attention. The 2D planar structure offers unique features such as high surface area, abundant surface active sites, efficient charge separation, and facile formation of heterostructures. The design and synthesis of 2D polymer nanosheets have greatly advanced the research in photocatalytic overall water splitting.

The recent advances in developing photocatalysts based on 2D polymer nanosheets for photocatalytic overall water splitting have been highlighted (11).

Synthetic polymers for photocatalytic water splitting were first demonstrated in 1985 (12). Poly-*p*-phenylene synthesized via a Ni-catalyzed cross-coupling reaction from 1,4-dibromobenzene was the first conjugated polymer that was able to generate H_2 using amines as sacrificial electron donors under ultraviolet light irradiation. Afterwards, various polymer photocatalysts with increasingly complicated structures and morphologies have been developed (13–17).

Recently, it has been demonstrated that 1,3-diyne-linked conjugated polymer nanosheets prepared by oxidative coupling of terminal alkynes such as 1,3,5-tris-(4-ethynylphenyl)benzene (TEPB) and 1,3,5-triethynylbenzene (TEB), c.f. Figure 3.8, exhibited unique capability in overall water splitting under visible-light irradiation.

Also, van der Waals heterostructures using ultrathin aza-CMP (conjugated microporous polymers) and C_2N nanosheets as O_2-evolving and H_2-evolving photocatalysts, respectively, were developed (18).

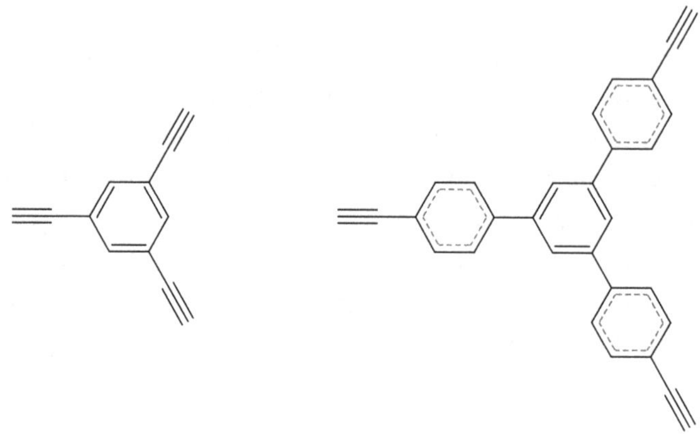

1,3,5-Triethynylbenzene 1,3,5-Tris(4-ethynylphenyl)benzene

Figure 3.8 Terminal alkynes.

3.3 Cleaning Methods

3.3.1 *Water Purification*

Water contamination and its purification are a global problem. The current approach to purify water is the reduction of the impurities to acceptable levels. One of the ways to achieve this is by using water-soluble polymers that extract organic and metallic contaminants from water (19). The use of chitin derivatives and chitosan derivatives for the detoxification of water and wastewater has been reviewed (20). These materials are effective biosorbents due to their high contents of amino and hydroxyl functional groups, which show significant adsorption potential for the removal of various aquatic pollutants.

Recently, a blend of composite polymers has been presented that eliminates both the contaminants simultaneously by the principle of adsorption at lower critical solution temperature (19). These composite polymers have been synthesized by grafting poly(*N*,*N*-diethylacrylamide), poly(*N*-isopropylacrylamide) and poly(*N*-vinylcaprolactam) onto the natural polymer chitosan or its derivatives, giving smart graft polymeric assemblies (GPAs). The monomers for the composite polymers are shown in Figure 3.9.

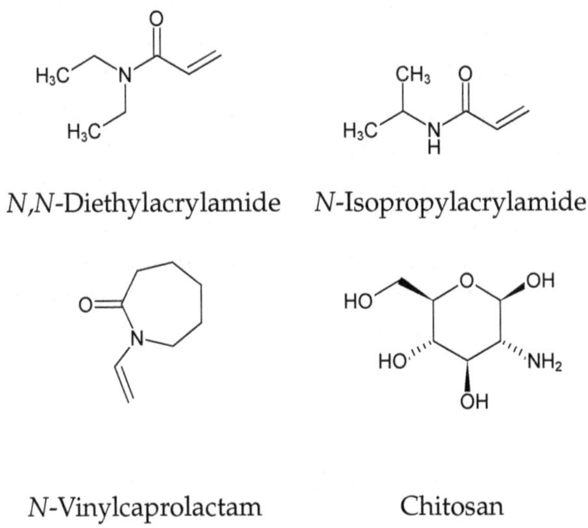

Figure 3.9 Materials for grafting.

One of the graft polymers, GPA-2, which contains *N,N*-diethylacrylamide, exhibits excellent adsorption properties able to remove metal ions like cadmium, cobalt, copper, lead, iron and also organic impurities like chlorophenol and phthalic anhydride. Studies reveal that 6 $mg\,ml^{-1}$ GPA-2 is able to effect 100% removal of the organic impurities – chlorophenol (50 *ppm*) and phthalic anhydride (70 *ppm*) – from water, while complete removal of the heavy metal ions (Cu^{+2}, Co^{+2} and Cd^{+2}) together at 30 *ppm* concentration has been achieved with 7.5 $mg\,ml^{-1}$ GPA-2. The reduction in level of impurities along with recyclability and reproducibility in the elimination spectrum makes these assemblies promising materials in water treatment (19).

Common impurities in effluents from industries include organic compounds like chlorophenols, benzopyrenes, polyaromatic hydrocarbons (PAHs), alkylphenols, phthalate esters, etc. (19). 2-Chlorophenol and phthalic anhydride were selected for studying the ability of these copolymers to adsorb organic impurities from water. The adsorption of the impurities was evaluated by UV-195 visible spectroscopy. Two different concentrations of chlorophenol with absorbance in the linear range of Beer-Lambert law (30 *ppm* and 50 *ppm*) were selected and these solutions were treated with the graft polymer assemblies. Each polymer of about 10 *mg* was dissolved in each of the selected concentration of chlorophenol and the solutions heated above the lower critical solution temperature of the polymer for 30 *min*. The solutions were then filtered to remove the precipitated polymer and the UV absorbance of the final solution was then measured at 273 *nm*. Similarly, 40 *ppm* and 70 *ppm* solutions of phthalic anhydride were treated with 10 *mg* of the graft polymer assemblies and the UV absorbance measured at 284 *nm*.

It was also of interest to test if the polymers have any preferential adsorption of one impurity in the presence of other impurities. To gauge the adsorption potential for impurities present simultaneously, an HPLC method was adopted using an Agilent Zorbax column and a Jasco PU-2080 binary pump system to determine the amount of impurities extracted by the polymers. The mobile phase used for analysis was methanol:water (45:55), pH 3.3 adjusted with 0.05% phosphoric acid and the wavelength used for detection was 256 *nm* (19).

The thermoresponsive polymers have been shown to be efficient in removal of organic compounds from water. On the other hand, the natural polymer chitosan has the capacity to remove inorganic ions and various dyes from water (19).

The adsorption of inorganic impurities of the composite polymer assembly GPA-2 occurs due to the unmasking of the hydrophilic chitosan moiety at the lower critical solution temperature, which naturally has a higher affinity for metal ions.

3.3.2 *Ammonia Capture*

A hypercrosslinked porous organic polymer was modified by post-oxidation and post-sulfonation to obtain a porous platform with a high density of acidic groups (21). Such an acidified material exhibits a record high NH_3 adsorption capacity per surface area, fast adsorption rate, and recyclability at low desorption temperature.

A dark-brownish solid was obtained from a microwave-assisted solvothermal reaction of toluene, formaldehyde dimethyl acetal, c.f. Figure 3.10, 1,2-dichloroethane, and anhydrous $FeCl_3$ as the catalyst. The application of the microwave method to the reaction system significantly reduced the reaction time to 5 *h*, compared with conventional reflux reactions with 18 *h* to 24 *h*. This can be ascribed to the fast nucleation and crystal growth under microwave irradiation.

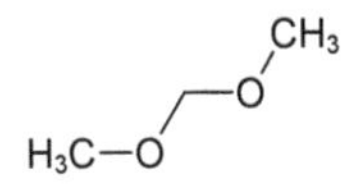

Figure 3.10 Formaldehyde dimethyl acetal.

In order to oxidize methyl groups on the benzene rings to carboxylic acid groups, the compound was suspended over a mixture of water and ethanol, and refluxed for 2 *d* after the addition of $KMnO_4$ and NaOH. The structure of such a polymer is shown in Figure 3.11. Then the material was treated with chlorosulfonic acid to give sulfonated polymer groups.

Furthermore, coating of the polymer with a hydroxyl-terminated poly(dimethylsiloxane) is a facile and efficient route to enable both a significant improvement of the low-pressure NH_3 adsorption capacity with a 40-fold enhancement from 0.04 $mmol\ g^{-1}$ to 1.41 $mmol\ g^{-1}$ with respect to the non-modified polymer at 500 *ppm* and hydrophobicity associated with the selective sorption of NH_3 over water vapor, which is hydrophilic for the non-coated material. This material is easy to prepare, cost-effective, and scalable to mass production (21).

3.3.3 *Removal of Heavy Metal*

3.3.3.1 *Nanofiber Membrane*

A hydrophilic poly(vinyl alcohol) (PVA)-*co*-ethylene nanofiber membrane for heavy metal ion removal was fabricated by the solid-phase synthesis of iminodiacetic acid (IDA), c.f. Figure 3.15, on nanofiber membrane surfaces (22). The hydrophilic PVA-*co*-ethylene nanofiber membranes could be activated with cyanuric chloride. The IDA was then covalently linked to the activated PVA-*co*-ethylene nanofiber membranes.

The chemical structures of activated and functionalized PVA-*co*-ethylene nanofiber membranes were confirmed with FTIR spectroscopy-ATR. The morphology of the PVA-*co*-ethylene nanofiber

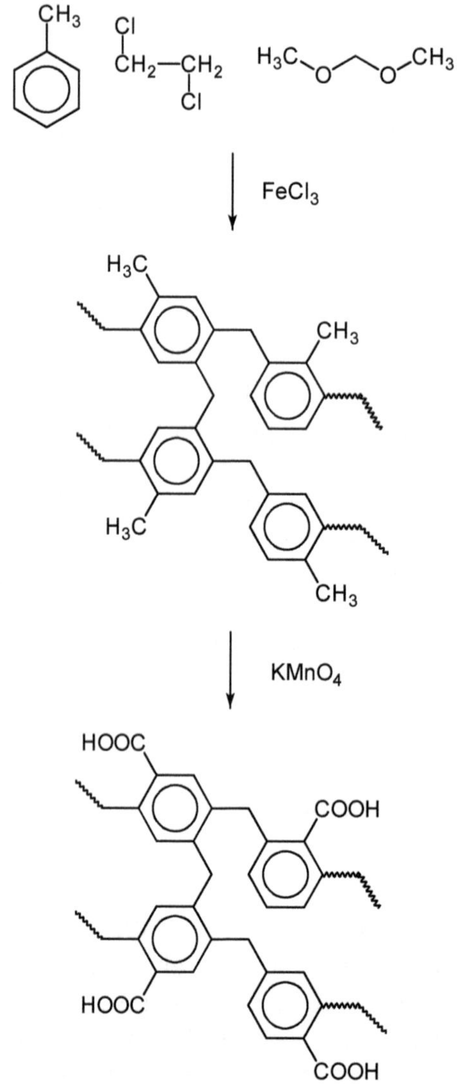

Figure 3.11 Synthesis of a hypercrosslinked polymer (21).

membranes was characterized with scanning electron microscopy (SEM).

The increase in the amount of IDA on functionalized PVA-*co*-ethylene nanofiber membranes significantly improved the adsorption amount of Cu^{2+}. The IDA functionalized PVA-*co*-ethylene nanofiber membranes demonstrated excellent adsorption capability of Cu^{2+}, Co^{2+}, Zn^{2+}, and Ni^{2+}. The adsorption of the above heavy metal ions could be repeatedly regenerated by desorbing the ions adsorbed on nanofiber membranes.

Thus, the IDA functionalized PVA-*co*-ethylene nanofiber membranes have great potential in applications for industry and drinking water treatment (22).

3.3.3.2 *Chromium(III) Ion-Imprinted Polymers*

Chromium(III) ion-imprinted polymers intended for the selective removal of Cr(III) ions from aqueous media were prepared using a Cr(III)-1,10-phenanthroline complex as a template by bulk polymerization (23). 1,10-Phenanthroline is shown in Figure 3.12.

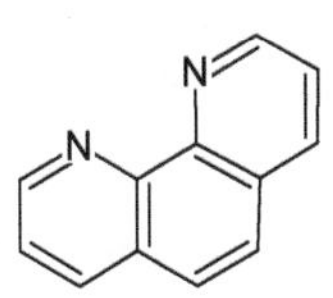

Figure 3.12 1,10-Phenanthroline.

The properties of ion-imprinted polymers synthesized using different functional monomers, including acidic (methacrylic acid) and neutral (styrene) agents, polar protic and aprotic porogens (ethanol, acetonitrile) and initiators (2,2′-azobisisobutyronitrile (AIBN), lauroyl peroxide, were studied in terms of their morphology and sorption properties towards Cr(III) ions, selectivity and reusability.

The studies indicated that a neutral functional monomer, aprotic solvent, and an azo initiator provided the best selectivity and sorption properties of the ion-imprinted polymer. Under optimal conditions, Cr(III)-ion-imprinted polymer prepared using styrene and AIBN showed good selectivity, reusability and stability. The selectivity coefficient was 16.7, 9.6 and 4.4 for Cr(III)/Cu(II), Cr-(III)/Fe(III) and Cr(III)/Mn(II), respectively. A good reproducibility of the results was obtained even after a hundred cycles. This testifies to the long-term stability of the sorbents. The ion-imprinted polymer made from styrene with AIBN can successfully be used as a sorbent in solid-phase extraction for selective chromium separation from the water solution (23).

3.3.3.3 *Heavy Metal Ions*

Low-cost polymeric adsorbents have been obtained in a simple reaction of diethylenetriamine (DETA) or pentaethylenehexamine (PEHA) with 1,3,5-tris(6-isocyanatohexyl)-1,3,5-triazinane-2,4,6-trione (HDI-IC). The monomers are shown in Figure 3.13.

The reagents used for the synthesis are inexpensive and are produced yearly in high quantities to be applied in various industrial processes. The resulting HDI-IC-DETA and HDI-IC-PEHA polymers possess ethyleneamine chains of various lengths. Therefore they are capable of forming poly(amine) chelating complexes with heavy metal ions. The composition, properties and morphology of the two polymers were investigated by FTIR spectroscopy, elemental analysis, thermogravimetry (TG), solid-state NMR, and SEM.

The influence of parameters such as pH, initial concentration, contact time, and temperature on the adsorption of Cd^{2+}, Co^{2+}, Cr^{3+}, Cu^{2+}, and Ni^{2+} ions was investigated. The adsorption isotherm data were well fitted by a Langmuir model, whereas the adsorption kinetics followed the pseudo-second-order kinetic model. The values of the maximum amount of ions adsorbed, calculated using the Langmuir model for adsorption of Cd^{2+}, Co^{2+}, Cr^{3+}, and Cu^{2+} ions on HDI-IC-PEHA polymer, were at least twice as high as those calculated for the HDI-IC-DETA polymer. The findings indicate that HDI-IC-DETA and HDI-IC-PEHA polymers are excellent adsorbents for the efficient removal of heavy metal ions from aqueous solutions (24).

Recovery of Gold Ions. Lignocellulosic coconut pith was functionalized with 3-aminopropyl-triethoxysilane, c.f. Figure 3.14, for preparing a aminopropyltriethoxysilane-functionalized lignocellu-

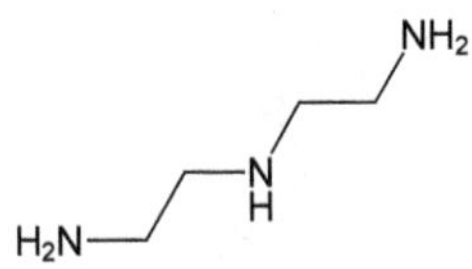

Diethylenetriamine

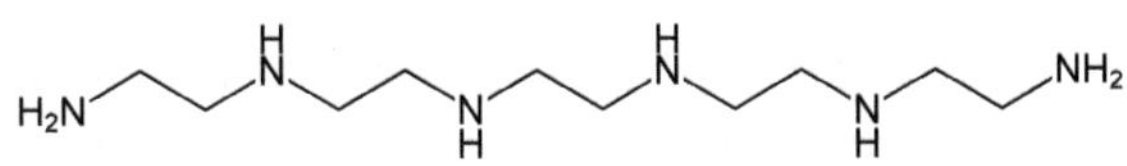

Pentaethylenehexamine

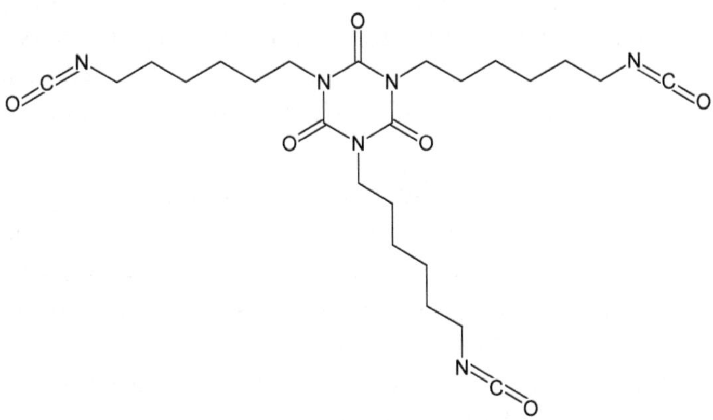

1,3,5-Tris(6-isocyanatohexyl)-1,3,5-triazinane-2,4,6-trione

Figure 3.13 Monomers for polymeric adsorbents (24).

losic coconut pith adsorbents for high adsorption affinity towards Au^{3+} ions (25).

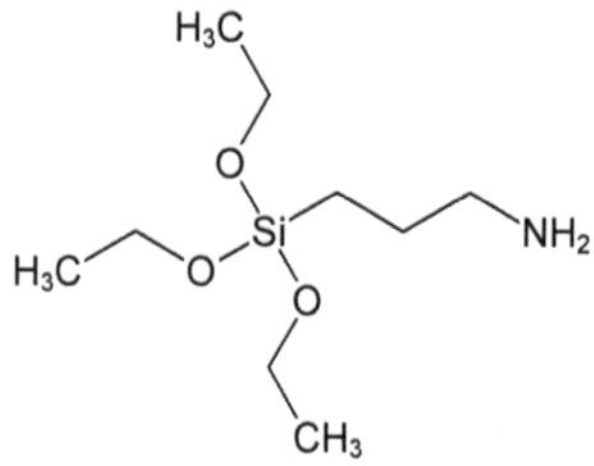

Figure 3.14 3-Aminopropyl-triethoxysilane.

The results of Au^{3+} adsorption showed that the 3-aminopropyl-triethoxysilane-lignocellulosic coconut pith possessed a much better Au(III) adsorption than lignocellulosic coconut pith and both exhibited an endothermic process.

The Au^{3+} adsorption isotherm data fitted well to the Langmuir isotherm having the maximum adsorption capacity of 215.68 $mg\,g^{-1}$ and 261.36 $mg\,g^{-1}$, respectively, for the lignocellulosic coconut pith and 3-aminopropyl-triethoxysilane-lignocellulosic coconut pith at 30°C (25).

A kinetic model analysis showed that the overall adsorption process was controlled by film diffusion, while the active site chemical interactions, e.g., ion exchange, chelation and reduction, were best described with a pseudo-second-order kinetic model. An adsorption-desorption experiment revealed that the 3-aminopropyl-triethoxysilane-lignocellulosic coconut pith could be regenerated with minimum loss of its adsorption capacity.

These results demonstrated the potential application of lignocellulosic materials as adsorbents through an appropriate modification for adsorptive recovery of Au^{3+} ions from an aqueous solution (25).

3.3.3.4 Functionalized D301 Resin

High adsorption capacity, fast adsorption rate, easy regeneration and good reusability were very important for qualified adsorbents used in removing toxic heavy metals from wastewater. A novel adsorbent was well designed and synthesized by functionalizing a D301 resin with IDA (26). IDA is shown in Figure 3.15

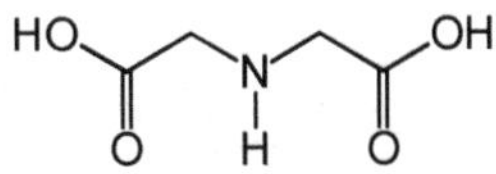

Figure 3.15 Iminodiacetic acid.

The physicochemical characteristics of IDA-functionalized D301 (ID301) were characterized by SEM, Brunauer-Emmett-Teller surface area analysis, FTIR and elemental analysis.

The adsorption performances of ID301 towards toxic heavy metal ions were systematically performed from the kinetics to isotherms and thermodynamics by a batch technique. The effects of contact time, initial metal concentration, pH, temperature and adsorbent dosage on adsorption performance were investigated. ID301 possesses a strong adsorption ability for Cu^{2+}, Cd^{2+}, and Pb^{2+}.

The pH and temperature have a great influence on the adsorption capacity. The adsorption capacities of ID301 towards Cu^{2+}, Pb^{2+}, and Cd^{2+} could reach 4.48, 2.99 and 2.26 $mmol\, g^{-1}$ at 293 K and pH of 5, respectively.

An adsorption thermodynamic experiment indicated that adsorption of ID301 towards the above-mentioned ions is an endothermic and spontaneous chemisorption process driven by entropy. In addition, ID301 could be reused without losing its adsorption capacity significantly (26).

3.3.3.5 *Transition Metals*

The synthesis of a hydrogel adsorbent for high selectivity and strong chelate removal of transition metals Cr, Cu and Ni was performed via a sol-gel method using maleic acid and (2,2′-ethylenedioxy)bis-(ethylamine) as ligand monomers (27).

A 14-membered macrocyclic functional unit was obtained, i.e., poly(1,4-dioxa-7,12-diazacyclotetradecane-8,11-dione) as the main active sites for metal adsorption. This compound is shown in Figure 3.16.

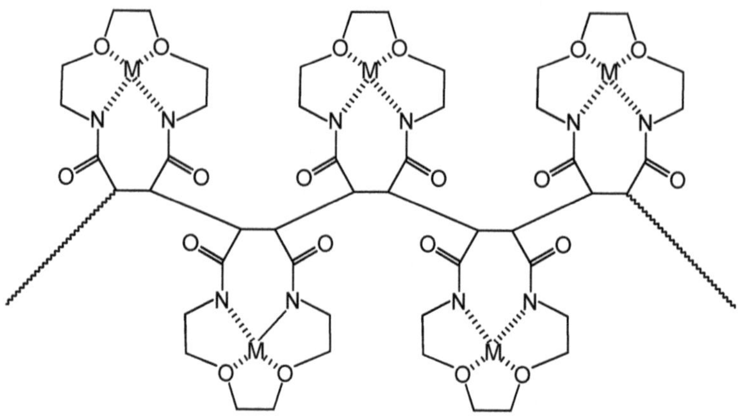

Figure 3.16 Poly(1,4-dioxa-7,12-diazacyclotetradecane-8,11-dione).

The crosslinked polymer chains of the adsorbent were clearly visualized using SEM spectroscopy, whereas the FTIR, 13C NMR and energy-dispersive X-ray spectroscopy techniques demonstrated its structural and functional groups.

Due to an improved chelating power, this macrocyclic hydrogel was able to ignore non-target substrates, instead showing high specificity for only Cu, Cr and Ni from both single and multi-ion competitive aqueous solutions, in the order Cr > Cu > Ni.

A pseudo-second-order kinetic model described the adsorption process, revealing chemisorption as the main mechanism which proceeded based on the host-guest chelation principle of metal ions onto the macrocyclic active sites of the gel. Thus, a high performance, high selectivity adsorbent system was achieved using active sites bearing substrate recognition and isolation properties (27).

3.4 Molecularly Imprinted Polymers

Functional monomers play a key role in preparing molecularly imprinted polymers by forming complex with templates to create recognition sites in the polymers (28). A new strategy was proposed to design functional monomers for efficient molecularly imprinted polymer synthesis.

In the propranolol imprinting process, methacrylic acid has always been used as a functional monomer, due to the efficient hydrogen bonding interactions between the carboxyl group in methacrylic acid and the 2-hydroxylethylamine group in propranolol. It was assumed that a functional monomer having a 2-hydroxylethylamine moiety may be used to imprint carboxylic acid molecules, e.g., naproxen. This compound is shown in Figure 3.17.

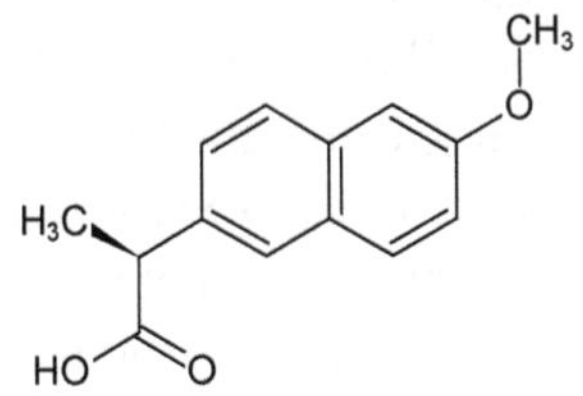

Figure 3.17 Naproxen.

To demonstrate this idea, a new monomer, 2-hydroxy-3-(isopropylamino)propyl methacrylate, was designed, c.f. Figure 3.18.

Computation results, by means of density functional theory method, revealed that 2-hydroxy-3-(isopropylamino)propyl methacrylate could form a stable complex with naproxen through hydrogen bonding interactions with the carboxylic acid group.

2-Hydroxy-3-(isopropylamino)propyl methacrylate was then used to synthesize naproxen-imprinted polymers using a precipitation polymerization method. Binding experiments showed that all the molecularly imprinted polymers could selectively recognize naproxen, confirming the feasibility of a paradigm shift in functional monomer design.

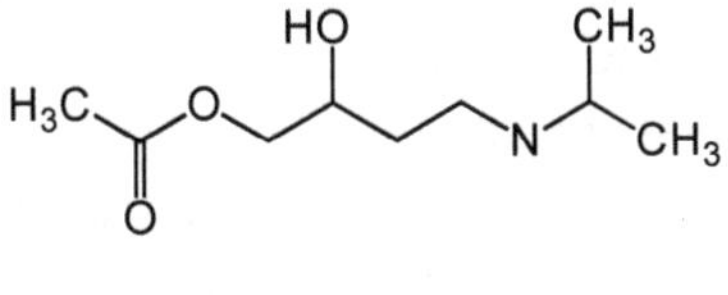

Figure 3.18 2-Hydroxy-3-(isopropylamino)propyl methacrylate.

The functional monomer 2-hydroxy-3-(isopropylamino)propyl methacrylate is a promising ligand that may be used to imprint other molecules having carboxylic acid or phosphoric acid groups. The paradigm shift in this study thereby opens a new avenue to design functional monomers for developing molecularly imprinted polymers (28).

3.5 Metal-Organic Frameworks

Metal-organic frameworks have received much attention because of their attractive properties. They show great potential applications in many fields. An emerging trend in the research of metal-organic frameworks is hybridization with flexible materials. This issue has been reviewed (29). The hybridization of metal-organic frameworks and polymers produces new and versatile materials that exhibit peculiar properties that are hard to realize with individual components.

Nanostructured materials, such as porous metal oxides, metal nanoparticles, porous carbons, and their composites, have been intensively studied due to their applications, including energy conversion and storage devices, catalysis, and gas storage (30).

Appropriate precursors and synthetic methods have been chosen for synthesizing such target materials. Metal-organic frameworks and coordination polymers have emerged as precursors to such nanomaterials, because they contain both organic and inorganic species.

Thermal conversion of metal-organic frameworks is a promising method for synthesizing functional nanomaterials that are difficult to obtain by conventional methods.

The usage of metal-organic frameworks/coordination polymers as precursors has been reviewed (30). Also, their transformation into functional nanomaterials with regard to the relationship between the intrinsic nature of the parent metal-organic frameworks and the daughter nanomaterials has been discussed.

For metal-based nanomaterials that are transformed from metal-organic frameworks, the nature of the metal ions in the metal-organic framework scaffolds affects the physicochemical properties. These are the phase, composite, and morphology of nanomaterials. Organic ligands are also involved in the chemical reactions with metal species during thermal conversion.

Along with the metal species, carbon is a major element in metal-organic frameworks. So, the appropriate choice of the precursor metal-organic frameworks and heat treatment can be expected to yield carbon-based nanomaterials (30).

3.6 Functional Microcapsules

A microfluidic approach for the fabrication of functional microcapsules via thiol-ene photopolymerization has been described (31).

An enhanced retention was shown of an encapsulated model small active (fluorescein) that was previously shown to be highly permeable (32). Fluorescein is shown in Figure 3.19.

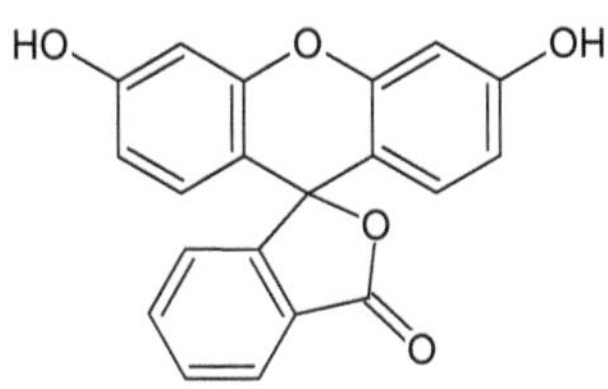

Figure 3.19 Fluorescein.

By inserting degradable anhydride monomers into the thiol-ene backbone, a tailored release kinetics could be demonstrated from a homogeneous degradable network, which aligns with the recent Microbead-Free Waters Act of 2015 (33). There, all nondegradable plastic particles less than 5 *mm* are to be banned for sale in the United States.

Also, thiol-ene microcapsules can be oxidized to tune the thermal-mechanical properties. In the study, trimethylolpropane tris-(3-mercaptopropionate), and either triethylene glycol divinyl ether, triallyl-1,3,5-triazine-2,4,6-(1H,3H,5H)-trione, or 4-pentenoic anhydride were used as monomers.

Furthermore, the rapid cure kinetics was shown by the thiol-ene chemistry in a continuous flow photopatterning device for hemispherical microparticle production.

To fabricate thiol-ene double emulsion drops, a glass capillary microfluidic device was used that contained two tapered cylindrical

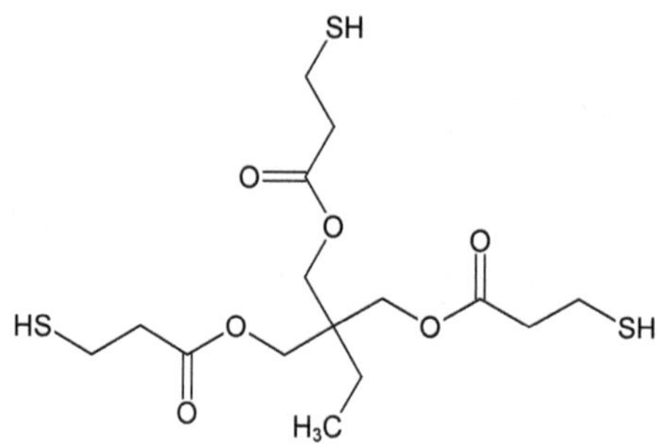

Trimethylolpropane tris(3-mercaptopropionate)

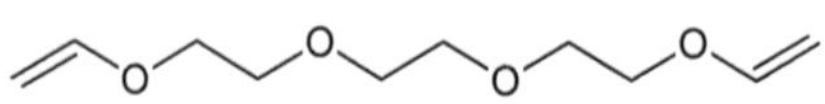

Triethylene glycol divinyl ether

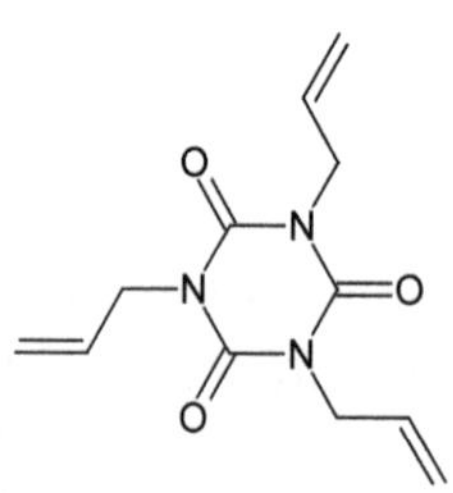

Triallyl-1,3,5-triazine-2,4,6-(1H,3H,5H)-trione

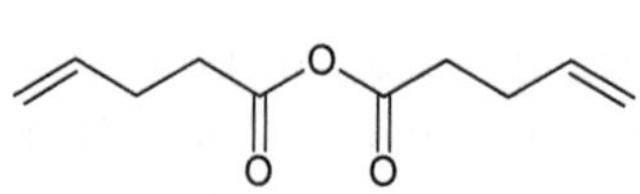

4-Pentenoic anhydride

Figure 3.20 Thiol monomers.

glass capillaries: One for the injection, and the other for collection (31).

Also, the ability of thiol-ene networks to be postfunctionalized was explored. A simple incubation of the microcapsules with H_2O_2 was done and an oxidation reaction was observed (31).

3.7 Shape-Memory Polymers

A hyperbranched poly(amine-ester) was synthesized from pentaerythritol tetraacrylate and diethanolamine, c.f. Figure 3.21, by a Michael addition reaction (34).

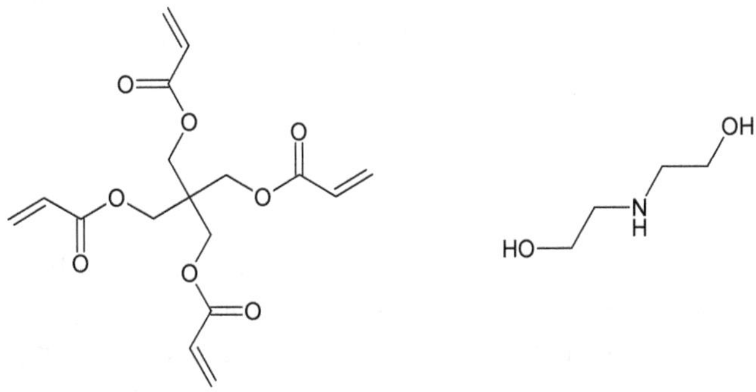

Pentaerythritol tetraacrylate Diethanolamine

Figure 3.21 Compounds for Michael addition.

A one-to-one stoichiometric reaction between diisocyanatodicyclohexylmethane and 2-hydroxyethyl acrylate produced dimers carrying both NCO and vinyl groups at two chain termini, which were subsequently reacted with hyper-OH to form hyperbranched polymers (HBP, Hyper-8). The replacement of 2-hydroxyethyl acrylate by trimethylolpropane diallyl ether, c.f. Figure 3.22, produced Hyper-16.

On the other hand, poly(urethane) (PU) prepolymers were synthesized from diisocyanatodicyclohexylmethane and polyol, end capped with 1,2-ethanedithiol, and UV cured to synthesize crosslinked PUs via a thiol-ene click chemistry. The hyperbranched

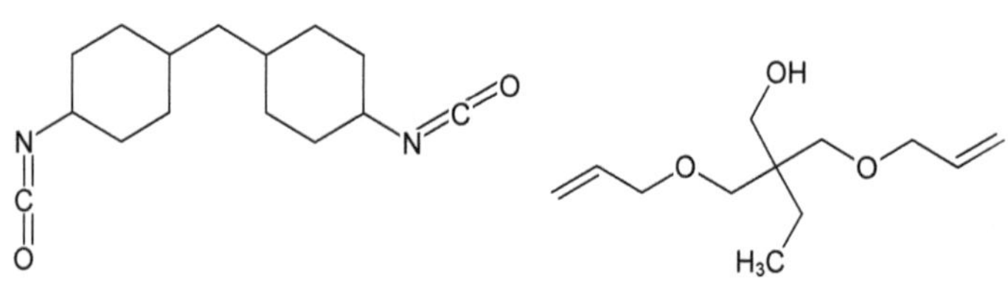

Diisocyanatodicyclohexylmethane Trimethylolpropane diallyl ether

Figure 3.22 Compounds for Michael addition.

polymers acted as multifunctional crosslinkers as well as reinforcing fillers and significantly enhanced the mechanical, thermal and shape memory properties. These effects were more pronounced with thiol-ene click chemistry than ene-ene curing (34).

3.8 Solder Pastes

Solder is widely used in the assembly of semiconductor packages and semiconductor devices (35). For example, solder balls or spheres are used in the assembly of semiconductor packages, such as in flip chip applications. It is known to place a stearic acid coating on the surface of such solder balls or spheres.

Solder paste is commonly used for surface-mounted soldering of electrical components to circuit boards. Solder paste is useful because it can be applied to selected areas of the circuit board with its tackiness characteristic providing the capability of holding the electrical components in position without additional adhesives before forming the permanent bonds as the board passes through the solder reflow process.

Solder paste typically comprises a solder powder, a resinous component such as rosin, activators such as organic acids or amines, rheological control agents, thickeners and solvents. The solder paste is typically coated on the circuit board by techniques such as screen printing, dispensing, and transfer printing.

Thereafter, the electrical components are placed on the circuit board and the solder paste is reflowed, by which the solder is heated sufficiently to cause it to melt, and thereafter cools the solder sufficiently to cause it to solidify.

One problem associated with the use of solder paste is that it often has a short and unpredictable shelf life, e.g., typically from about one month to six months (35). The unpredictability in shelf life is caused, at least in part, by variations in the lag time from when the solder powder is made to the time it is mixed with flux to form solder paste, thereby resulting in variations in the degree of oxidation on the solder powder.

Such oxidized powder does not reflow as well as unoxidized powder. Furthermore, when the solder powder is combined with flux, which is inherently corrosive, the solder powder often reacts with the flux, thereby oxidizing the powder and reducing the acidity, thus effectiveness, of the flux. As a result, the performance of the solder paste often deteriorates over time. Moreover, the reaction between the solder powder and the flux typically causes the viscosity of the solder paste to increase substantially, which can make printing the solder paste difficult, if not impossible, depending on the pitch.

Attempts have been made to reduce the reaction rate between the solder powder and the flux and thereby increase the shelf life of the solder paste, by storing the solder paste under refrigeration conditions. However, refrigeration is not effective to compensate for the varying degrees of oxidation on the solder powder prior to its incorporation into the solder paste.

A metal powder has been developed that is having an organic acid or latent organic acid coating on at least a portion of a surface thereof, where the coated metal powder is particularly suitable for use in solder pastes (35).

The organic acid or latent organic acid-functionalized polymer for use as the coating has as its chief function the task of physically isolating the metal particles from environmental degradation, such as oxidation and chemical reaction with flux media. In general, the organic acid or latent organic acid coating acts as a physical barrier toward oxidation while the metal powder and/or solder paste in which the coated metal particle is being stored for use.

A method of forming a polymer coating on a metal particle includes the steps of (35):

1. Providing a plurality of metal particles,

2. Applying to the plurality of metal particles a polymer having an organic acid or latent organic acid functional group under suitable conditions so as to substantially coat at least a portion of the surface of most of the metal particles, and
3. Exposing the organic acid- or latent organic acid-functionalized polymer coating on the surface of the metal particles to suitable conditions to form an acetal or hemi-acetal linkage between the organic acid- or latent organic acid-functionalized polymer and the surface of the metal particle.

Several poly(butadiene) and maleic anhydride adducts were synthesized with different maleic anhydride contents ranging from 5% to 0.1% by weight, and were evaluated. The synthesis is described below. The synthesis procedure of a latent organic acid is (35):

Preparation 3–2: First, 200 *g* of poly(butadiene) (MW 3400) and maleic anhydride (1.0 g) were dissolved in 250 *ml* of toluene; this mixture was transferred into a Parr pressure vessel and was heated to 220°C to 250°C on a hotplate. This enclosed system was allowed to stand at this temperature for a period of time of about 3–4 *h* and then allowed to cool to room temperature. The reaction mixture was diluted with 2 *l* of toluene and passed through a glass funnel with a thin layer of silica gel to remove the residual unreacted maleic anhydride. The solvent was then removed by rotary evaporation to give the product a viscous yellow liquid.

3.9 Antimicrobial Food Packaging Films

Food packaging is an important part of food products, both to protect food quality and safety of food products to enhance their added value. Food packaging materials with sufficient mechanical strength, barrier properties, thermal stability, biodegradability, and antibacterial and antioxidant properties are necessary for food safety and extending the shelf life of packaged foods.

Ternary blend films were prepared with different ratios of starch/PVA/citric acid. The solvent casting method was used for the preparation (36).

Preparation 3–3: Film solutions were prepared by dissolving 2.81 *g* of starch, PVA into 30 *ml* of distilled water with 2.11 *g* of glycerol as a plasticizer while mixing vigorously for about 45 *min* at 95°C using an electric stirrer. The starch/PVA and citric acid composite films were prepared by the solution-casting method. First, PVA was dissolved in distilled water at 95°C while the corn starch was gelatinized at 90°C. Then, citric acid was added to the PVA solution at 80°C and gelatinized starch and glycerin were added with stirring for 30 *min*. A transparent and uniform film fluid was obtained.

The amounts of the compounds in the prepared compositions and the baking times are collected in Table 3.1.

Table 3.1 Designs of the compositions (37).

PVA /[*g*]	Starch /[*g*]	Glycerol /[*g*]	Citric acid /[*g*]	Baking time /[*min*]
2.81	2.81	1.87	0	120
2.81	2.81	2.11	1	120
3.75	1.25	2.5	1	120
2.81	2.81	1.87	0	270
2.81	2.81	2.11	1	270
3.75	1.25	2.5	1	270
2.81	2.81	1.87	0	300
2.81	2.81	2.11	1	300
3.75	1.25	2.5	1	300

The influence of different ratios of starch/PVA/citric acid and different drying times on the performance properties, transparency, tensile strength, water vapor permeability, water solubility, color difference, and antimicrobial activity of the ternary blend films were investigated.

The films were highly transparent. The films with a ratio of starch/PVA/citric acid of 3:3:0.08 showed a 54.31 times water-holding capacity of its own weight and its mechanical tensile strength was 46.45 *MPa*. In addition, the surface had good uniformity and compactness. Films with a ratio of starch/PVA/citric acid of 3:1:0.08 and 3:3:0.08 showed strong antimicrobial activity to *Listeria monocytogenes* and *Escherichia coli*, which were the foodborne pathogenic bacteria used.

Freshness test results of fresh figs showed that all of the blends prevented the formation of condensed water on the surface of the film, and both films prevented the deterioration of figs during storage. The films can be used as an active food packaging system due to their strong antibacterial effect (37).

3.10 Flame Retardants

3.10.1 Flame-Retardant Epoxy Nanocomposite

9,10-Dihydro-9-oxa-10-phosphaphenanthrene-10-oxide (DOPO), c.f. Figure 3.23, has been grafted onto the surface of graphene oxide (GO) by reacting epoxy ring groups together with the reduced graphene structure, DOPO-rGO (38).

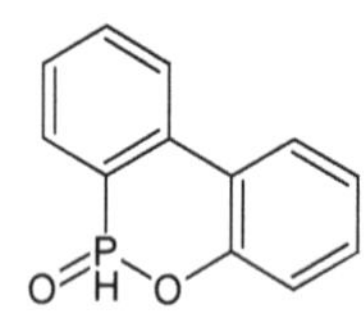

Figure 3.23 9,10-Dihydro-9-oxa-10-phosphaphenanthrene-10-oxide.

Several spectroscopic methods confirmed that DOPO not only covalently bonded to the GO as a functionalization moiety, but also partly restored the conjugate structure of GO as a reducing agent. A pellet-like structure of DOPO on rGO sheets was observed by means of transmission electron microscopy (TEM). This structure contributes to a good dispersion of rGO in nonpolar toluene.

Furthermore, the flame retardancy and thermal stability of DOPO-rGO/epoxy nanocomposites containing various weight fractions of DOPO-rGO were investigated by the limiting oxygen index (LOI) test and TG in nitrogen. Significant increases in the char yield and the LOI were achieved with the addition of 10% DOPO-rGO in the epoxy resin, giving improvements of 81% and 30%, respectively.

DOPO-rGO/epoxy nanocomposites with phosphorus and graphene layer structures were found to contribute to excellent flame retardancy compared to that of neat epoxy. Therefore, the synergistic effect of DOPO-rGO is quite useful. So, this material can be used as a potential flame retardant (38).

In another study, a related compound, 10-(2,5-dihydroxyl-phenyl)-9,10-dihydro-9-oxa-10-phosphaphenanthrene-10-oxide, c.f. Figure 3.24, was used to improve the dispersion and fire-retardant property in epoxy resins (39).

In order to improve the dispersion and fire-retardant property in epoxy resin (EP), GO was functionalized via surface modification by

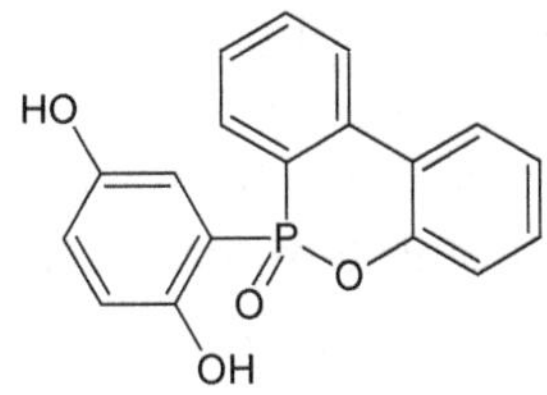

Figure 3.24 10-(2,5-Dihydroxyl-phenyl)-9,10-dihydro-9-oxa-10-phosphaphenanthrene-10-oxide.

a flame retardant which was synthesized by the reaction of methyl dichlorophosphate and 10-(2,5-dihydroxyl-phenyl)-9,10-dihydro-9-oxa-10-phosphaphenanthrene-10-oxide. Several compositions with different ratios of GO and functionalized GO epoxy nanocomposites were obtained by *in-situ* polymerization.

The incorporation of the functionalized GO effectively enhanced the thermal stability and flame retardancy of the epoxy nanocomposites. The thermal properties of the nanocomposites were investigated by TG experiments in nitrogen atmosphere. These experiments indicated that functionalized GO can improve the char residues. The flame retardancy of the nanocomposites was characterized by a cone calorimeter test. The results showed that the incorporation of 2% functionalized GO into EP decreased the value of peak heat release rate, total heat release, average effective heat of combustion, peak values of the CO release rate and the CO_2 release rate by 25%, 28%, 29.5%, 27% and 29%, respectively (39).

3.10.2 *Graphene Grafted Poly(phosphamide)*

A poly(phosphamide) (PPA) was synthesized and covalently grafted onto the surface of graphene nanosheets (GNSs) to obtain a novel flame retardant, PPA-g-GNS, and subsequently PPA-g-GNS was incorporated into EPs to enhance the fire resistance of the compositions (40).

The chemical structures and morphology of the precursors and target product were confirmed using ^{1}H-NMR spectroscopy, FTIR, and atomic force microscopy. The tensile results showed that the mechanical strength and modulus of the PPA-g-GNS/EP composite were higher than those of pure EP and PPA/EP, owing to the outstanding reinforced effect of graphene. The evaluation of the thermal properties demonstrated that the addition of PPA or PPA-g-GNS to epoxy had a thermal destabilization effect below 400°C, but led to a higher char yield at higher temperatures.

Also, the PPA-g-GNS/EP composite exhibited a superior fire resistant performance, such as higher LOI values, and reduced the peak heat release rate and fire index of growth rate values, compared to pure EP and PPA/EP. This was probably attributed to the higher char yield during combustion (40).

A possible flame-retardant mechanism was speculated according to the direct pyrolysis-mass spectrometry results: The phosphate species degraded from PPA catalyzed the decomposition of the PPA-g-GNS/EP composites to generate various pyrolysis products. The pyrolysis products were absorbed and propagated on the graphene which served as a template of micro-char, and thus a continuous and compact char layer was formed. Such a char layer provided effective shields to protect the underlying polymers against flame (40).

3.10.3 *Poly(propylene)/Multiwall Carbon Nanotube Nanocomposites*

Poly(propylene) carbon nanotube nanocomposites influenced by surface functionalization and surfactant molecular weights have been studied (41).

3-Aminopropyl-triethoxysilane, c.f. Figure 3.25, was utilized to modify the CNTs, and maleic anhydride grafted poly(propylene) (PP) with the molecular weights of 800 *Dalton* and 8000 *Dalton* was used to further improve the dispersion of the CNTs in the PP matrix.

TG and microscale combustion calorimetry revealed that the molecular weight of maleic anhydride grafted PP directly affects the thermal stability and flammability of the materials: Both maleic anhydride grafted PP polymers increase the thermal stability of PP. However, the heat release rate of PP/functionalized carbon nanotubes is reduced in the presence of maleic anhydride grafted PP of 800 *Dalton* and increased in the presence of maleic anhydride grafted PP of 8000 *Dalton* (41).

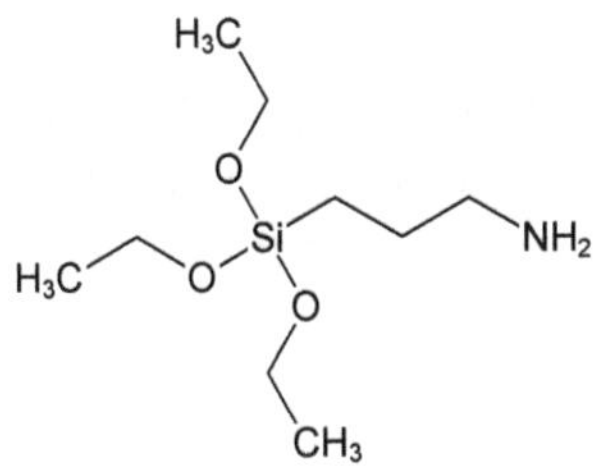

Figure 3.25 3-Aminopropyl-triethoxysilane.

3.10.4 *Poly(vinyl alcohol) Composites*

Surface functionalization of molybdenum disulfide (MoS_2) was prepared by a simple reflux reaction. Here, octa-vinyl polyhedral oligomeric silsesquioxane (POSS) was used (42).

The structure of octa-vinyl POSS-MoS_2 was confirmed by several analysis methods. The SEM and TEM results of fracture surface exhibited that octa-vinyl POSS-MoS_2 was well dispersed in the matrix due to the good interfacial interaction between the functionalized MoS_2 and PVA. TG and differential scanning calorimetry (DSC) indicated that the thermal decomposition temperature and the glass transition temperature were improved. Compared with pure PVA, the maximum degradation temperature of the synthesized composites was increased by 23°C, and the glass transition temperature was improved by 10.2°C.

The peak of heat release rate and the total heat release were decreased. The tensile stress was increased by 57% with addition of 2% octa-vinyl POSS-MoS_2. Moreover, the addition of octa-vinyl POSS-MoS_2 significantly decreased the gaseous products, including hydrocarbons, carbonyl compounds and carbon monoxide, which was attributed to the synergistic effect of octa-vinyl POSS and MoS_2. The adsorption and barrier effect of MoS_2 inhibited the heat and gas release and promoted the formation of graphitized carbons, while octa-vinyl POSS improved the thermal oxidative resistance of the char layer (42).

3.10.5 *Ethylene Vinyl Acetate Copolymer*

A phosphorus-nitrogen-containing compound, *N*-(2-(5,5-dimethyl-1,3,2-dioxaphosphinyl-2-ylamino)-hexylacetamide-2-propyl acid) (PAHPA) was synthesized and characterized (43). The synthesis of PAHPA is shown in Figure 3.26.

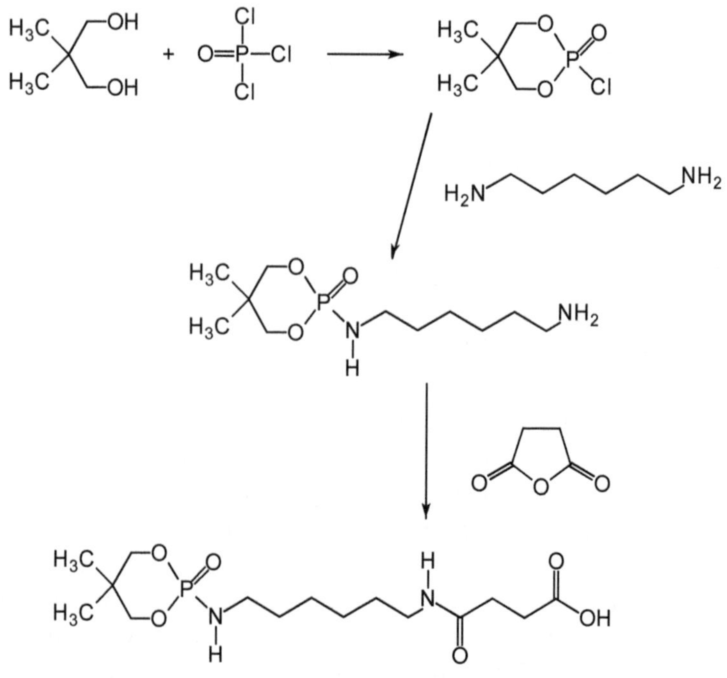

Figure 3.26 Synthesis of *N*-(2-(5,5-dimethyl-1,3,2-dioxaphosphinyl-2-ylamino)-hexylacetamide-2-propyl acid) (43).

A flame retardant, i.e., layered double hydroxides modified with PAHPA (PAHPA-layered double hydroxides), was prepared by ion exchange of layered double hydroxides with PAHPA.

The results from FTIR spectroscopy, X-ray photoelectron spectroscopy and energy-dispersive X-ray analysis with a high-angle annular dark-field scanning transmission electron microscope show that PAHPA intercalated layered double hydroxides. The X-ray diffraction and TEM results show that PAHPA-layered double hydroxides achieve good dispersion in ethylene vinyl acetate copolymer (ethylene-vinyl acetate (EVA)) matrix and the

EVA/PAHPA-layered double hydroxide nanocomposites, i.e., EVA filled with 5% PAHPA-layered double hydroxides, are formed by polymer melt intercalation.

Thermal stability and flammability properties were investigated by TG and cone calorimeter tests. The results showed that the addition of PAHPA-layered double hydroxides improves the thermal stability and reduces the flammability of the EVA resin. In comparison to a pure EVA resin, the peak heat release rate of the EVA/PAHPA-layered double hydroxide nanocomposites is reduced by about 43%. The results of SEM and TEM indicate that a compact and dense intumescent char is formed for the EVA/PAHPA-layered double hydroxide nanocomposites after combustion (43).

3.10.6 *Acrylonitrile-Butadiene-Styrene Polymers*

A acrylonitrile-butadiene-styrene (ABS)-*g*-glycidyl methacrylate powder functionalized with epoxy groups was prepared with poly(butadiene) as core and styrene-*co*-acrylonitrile-*co*-glycidyl methacrylate as shell. The ABS-*g*-GMA powder, styrene acrylonitrile copolymer and triphenyl phosphate (TPP) were melt blended together and halogen-free flame-retardant ABS resin was prepared (44).

It was found that when the content of TPP was 20%, the epoxy-functionalized ABS resin showed much higher thermostability than the ordinary ABS resin. The burning rate of the epoxy-functionalized resin decreased from 35 $mm\ min^{-1}$ to 8.6 $mm\ min^{-1}$, the mass loss rate decreased from 25% min^{-1} to 10% min^{-1}, and the LOI increased from 22% to 28%. FTIR experiments showed that the chemical reactions took place between epoxy-functionalized ABS resin and the decomposition products of TPP. The tensile tests and impact tests revealed that the epoxy-functionalized ABS/TPP samples showed a higher elastic modulus, tensile strength and impact strength than conventional ABS/TPP samples (44).

3.11 Liquid Toner

A liquid toner dispersion is a stabilized dispersion for use in a printing process (45). It differs from other ink dispersions such as those for offset printing and inkjet compositions, on the basis of the particles it contains.

Whereas an inkjet composition generally contains a pigment as such, the liquid toner dispersion contains marking particles with pigment that is mixed with or embedded in a polyester resin binder. The resulting particles have a suitable diameter in the range of 0.5 μm to 5.0 μm, whereas the pigment particle size in inkjet and offset is below 500 nm.

One of the complexities of a liquid toner process is the stability of the liquid toner dispersion. Therefore, a dispersant is added. The use of hyperdispersants seems beneficial. These hyperdispersants comprise an anchor group and a stabilizing group. The anchor group is anchored on the polymer particle surface by single-point or, typically, multipoint anchoring.

Both acrylates and amines are known as anchor groups. The stabilizing group grafted onto the anchor group extends in a non-aqueous system to provide steric stability. One specific example of a known hyperdispersant is a graft copolymer with a poly(ether imide) (PEI) as the anchor group and poly(12-hydroxy stearic acid) as the solvent group in aliphatic hydrocarbon continuous phases. This graft copolymer of a hydroxylated fatty acid is commercially available, for instance, from Lubrizol under the tradename Solsperse™. Other known hyperdispersants are available from Tianlong Chemicals under the tradename of TiloSperse™.

It has been found in relation to liquid toner dispersions that the stability requirements are manifold. The dispersion should evidently be stable during and after preparation, i.e., during storage and upon application to the first member. But the liquid toner process further requires that the fusing is not hindered or disturbed by the presence of the same dispersant. Moreover, the dispersion should not be sensitive to an artifact called caking, after charging the dispersion and/or discharging the dispersion and when the toner layer is mechanically stressed, e.g., when blade cleaning is performed.

In order to transfer the liquid toner dispersion from the first member to the second member, the liquid toner dispersion is typically charged. Due to this charging process, the transfer may be selective, such that merely a desired image is transferred from the first member to the second member. However, charging and compaction of the liquid toner dispersion also has an impact on the stability of the dispersion.

As a consequence, an issue may occur that is known as *caking*. Polymer marking particles in the dispersion particularly tend to form lumps in the dispersion, resulting in a liquid with a non-uniform distribution of the marking particles. This caking often results in an increase of the viscosity of the liquid dispersion. This viscosity increase is significant, and could be a tenfold increase or even more, resulting in a more difficult liquid system to be transported.

It has been thought that caking is the result of marking particles that come so close into each other's neighborhood on the developing member, that they start to feel each other's presence and start interacting with each other. Caking can also be the result of injecting charge and applying high shearing forces, which are typically present when a thin layer of liquid developer dispersion passes through a very narrow gap between two (rotating) members of the printing apparatus or huge (microsized) mechanical interaction like a cleaning blade scraping on a circular surface.

Therefore, there is some need to develop a liquid toner dispersion that would meet all the above-mentioned needs. The hyperdispersant should in particular be capable of stabilizing the initial liquid toner dispersion, but it should not disturb the fusing process and should not give rise to significant caking.

A liquid toner dispersion has been described that consists of a grafted copolymer of an amine-functionalized polymer onto which a fatty acid compound is grafted (45).

Examples of preferred amine-functionalized polymers are poly(amine)s, for instance poly(allylamine)s and poly(alkylene)imines, wherein the alkylene is chosen from ethylene, propylene, isopropylene, butylene, isobutylene and any other butylene isomer. The amine-functionalized polymer may also be a copolymer.

For example, the synthesis of poly(hydroxystearic acid) has been detailed. This is a polyester of 12-hydroxy stearic acid. The carboxyl group at the end of this compound is then reacted with pentaethylenehexamine to form an amide at the end (45). The monomers are shown in Figure 3.27.

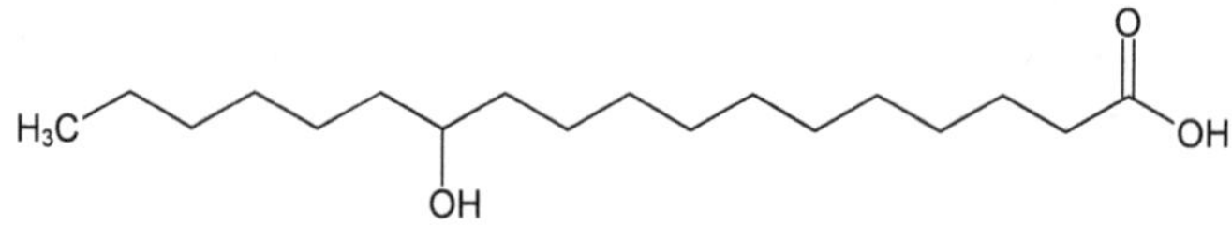

12-Hydroxy stearic acid

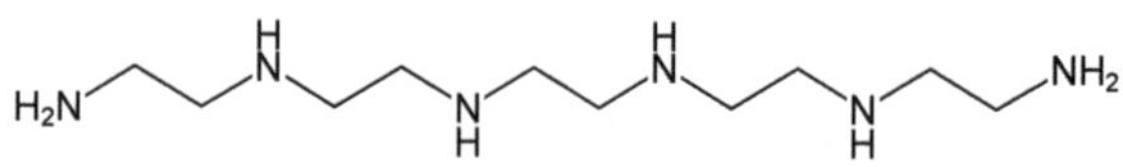

Pentaethylenehexamine

Figure 3.27 Monomers for poly(hydroxystearic acid) (45).

3.12 Hydroxyl-Functionalized Compositions

Copolymers containing keto groups as functional groups, also referred to as polyketones, are well-known molding and blending resins. For example, copolymers from ethylene and carbon monoxide are commonly used as plasticizers for solid organic polymers such as poly(vinyl chloride).

By a chemical reaction, the keto groups in polyketones can at least be partly converted into a variety of other functional groups. The chemical modification, changing the keto groups into other functional groups, changes the properties of the polyketones and renders them eligible for uses for which the original polymers were not or not very suitable. The reduction of polyketones produces polyalcohols, which can be used as adhesives or coatings in a number of applications.

Various methods have been used for the reduction of these polyketones to polyalcohols. The use of copper chromite or nickel catalysts to reduce polyketones to polyalcohols by hydrogenation using a transition metal catalyst has been documented (46). The starting materials are linear alternating polymers of carbon dioxide and at least one ethylenically unsaturated monomer. Terpolymers of carbon monoxide, ethylene, and propylene can also be employed.

The hydrogenation of polyketones using nickel salts reacted with borohydrides as a catalyst in an alcoholic media has been reported

(47). Also, an improved process for reducing polyketones containing from 1 to 50% of keto groups to polyalcohols using borohydride salts, where water was used instead of alcohols, has been described (48).

There, the reaction is conducted in a solution or suspension for an extended period of at least a few hours (49). This type of reaction necessitates the costly process of removing the solvent. Furthermore, the use of borohydrides releases hydrogen gas, potentially a safety hazard due to the inherent flammability of hydrogen.

Therefore, it would be desirable to conduct the reaction converting the keto group to polyalcohols in such a way that does not require the solvent recovery step. It would also be beneficial to use a reducing agent that does not pose a fire and explosion problem associated with hydride reducing agents.

It has been found that copolymers containing keto groups can be reduced to polyalcohols by using aluminum alkoxides (49). The reduction reaction of the polyketone polymer in molten condition with a metal alkoxide, preferably aluminum alkoxide, is carried out in the presence of a stoichiometric or excess amount of (49):

1. Isopropyl alcohol,
2. An alcohol which is substituted with a bulky alkyl group, or
3. Ketone which is substituted with a bulky alkyl group, which reacts with the metal alkoxide salt of the polymer to regenerate metal alkoxide, forming a polyalcohol.

An exemplary preparation runs as follows (49):

Preparation 3–4: In a HAAKE mixer, a copolymer containing 13% carbon monoxide and ethylene constituting the balance was prepared by the method of free-radical high-pressure polymerization. The polymer was added first and mixed at about 110°C and at a mixer setting of 200 *rpm* until molten. Aluminum isopropoxide was then added and the mixture was blended at a melt temperature of about 150°C. A rapid reaction was noted in about 5 to 10 *min* as the polymer mass became dry and fluffy. The melt temperature was raised to about 225°C and mixing was continued for about another 30 *s*.

3.13 Polymeric Membranes

3.13.1 *Proton Exchange Membranes*

A sulfonated polymer brush-functionalized graphene oxide (SPB-FGO) was prepared by reversible addition-fragmentation chain transfer polymerization. The composition was introduced into a sulfonated poly(ether sulfone) matrix to fabricate composite proton exchange membranes (50).

The method for synthesis of a sulfonated poly(ether sulfone) has been detailed (51). The synthesis route is shown in Figure 3.29.

Preparation 3–5: First, 1.0328 *g* 9,9′-bis(4-hydroxyphenyl)fluorene, 0.6314 *g* 4,4′-dihydroxybenzophenone, 1.4988 *g* 4,4′-sulfonylbis(fluorobenzene), 1.635 *g* potassium carbonate, 30 *ml* *N,N*-dimethylacetamide, and 15 *ml* toluene were added into a three-neck flask equipped with a magnetic stirrer, a Dean-Stark trap and a nitrogen inlet. The reaction was performed under reflux at 140°C for 4 *h* under nitrogen protection. Then the temperature of the reaction system was increased to 165°C and kept for 20 *h* while the toluene was removed from the system completely. As the reaction proceeded, the viscosity of the solution gradually increased. After the reaction, the system was poured into 500 *ml* hot ethanol and dried at 70°C under vacuum for 24 *h*.

The influence of the loading content of SPB-FGO on the properties of the composite membranes was studied (50). It could be demonstrated that proton conductivity, thermal stability, mechanical property and oxidative stability for the composite membranes, were improved in comparison to a pristine sulfonated poly(ether sulfone) membrane.

The methanol permeability of the composite membranes was lower than that of the pristine membranes. The proton conductivity of sulfonated poly(ether sulfone)/SPB-FGO-2 was nearly doubled as compared with that of the pristine membrane at 25°C. This result could be attributed to the fact that the SPB-FGO was well-distributed within the sulfonated poly(ether sulfone) membrane matrix and the ion clusters were relatively uniform in size, which made the SPB-FGO fillers create broad ionic pathways through the sulfonic acid groups in polymer brushes via interfacial interactions.

The study demonstrated that the incorporation of an appropriate amount of SPB-FGOs can improve the properties of proton exchange membranes (50).

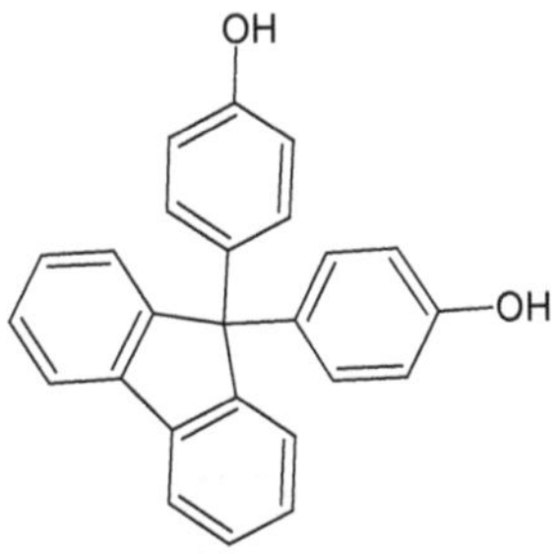

9,9′-Bis(4-hydroxyphenyl)fluorene

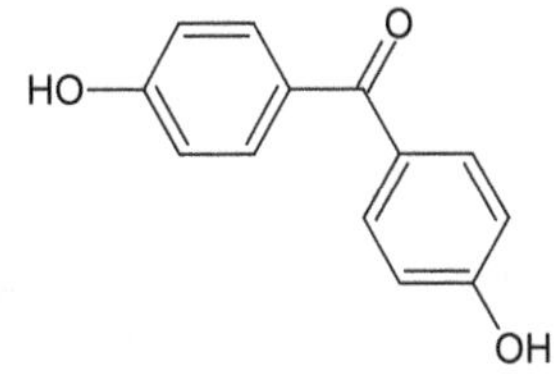

4,4′-Dihydroxybenzophenone

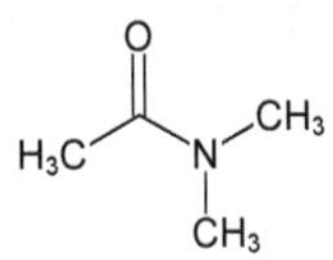

N,*N*-dimethylacetamide

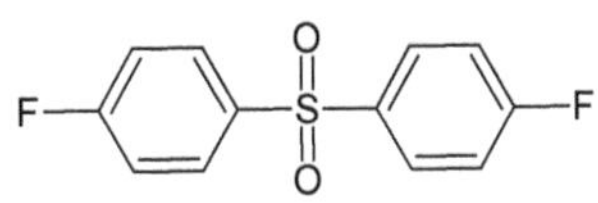

4,4′-Sulfonylbis(fluorobenzene)

Figure 3.28 Components for sulfonated poly(ether sulfone).

3.13.2 *Imidazole-Based Anion Exchange Membranes*

The recent development of alkaline stable imidazole-based anion exchange membrane and related ionomers has been reviewed (52). Dioxide Materials Inc. has developed a group of imidazole-functionalized membranes using a poly(styrene) (PS)-based backbone that are surprisingly stable in strong alkaline solutions (53–55).

Imidazolium-functionalized styrenic copolymers have been widely studied for their use as anion exchange membranes in carbon dioxide and water electrolysis systems (56). New developments in catalytic electrode materials suggested that electrolyzers based on this technology will be commercially viable.

Important aspects of the physicochemical behavior of these membrane materials, especially transformations occurring upon activation in strongly alkaline media, have been discussed. *N*-methylimidazolium-functional styrenic copolymer membrane materials recovered after activation by exposure to 1*M* KOH solution and reacidification in HCl (56).

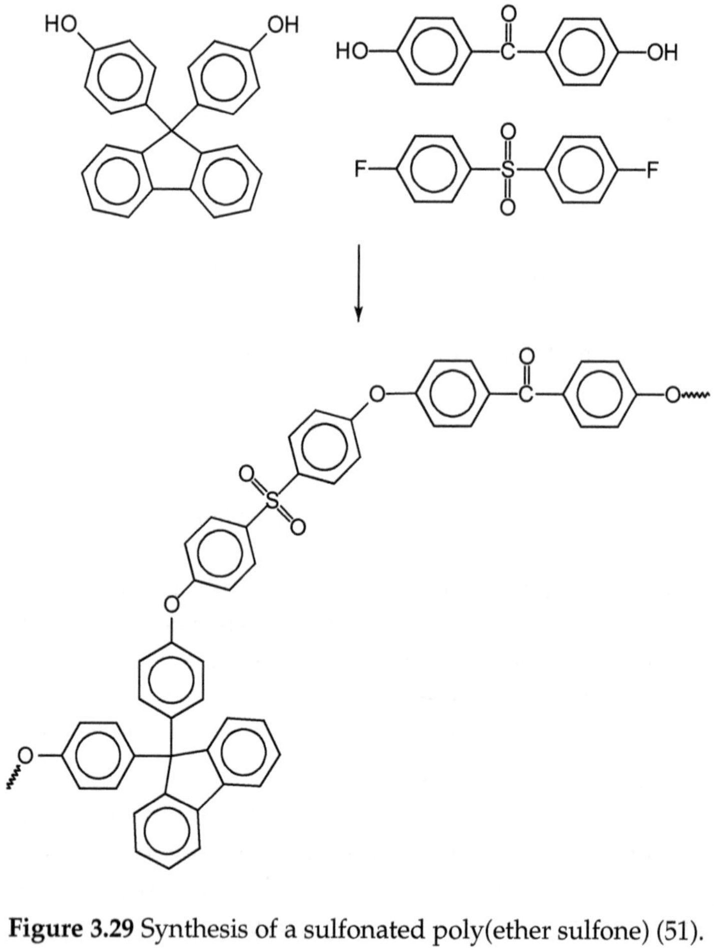

Figure 3.29 Synthesis of a sulfonated poly(ether sulfone) (51).

3.13.3 *Organic Solvent Nanofiltration*

A nanometric functionalized polymer of intrinsic microporosity (PIM), i.e., thioamide-PIM-1 (TPIM), has been synthesized and used for the removal of dyes from ethanol and acetone solutions (57).

The presence of thioamide groups in TPIM provides an anchor to further crosslink the polymer with trimesoyl chloride so that the resultant membranes not only have a tight pore size, i.e., a small free volume, but also have better chemical stability and rejection. Trimesoyl chloride is shown in Figure 3.30.

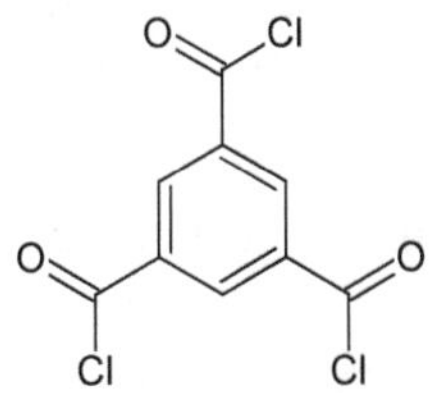

Figure 3.30 Trimesoyl chloride.

The organic solvent nanofiltration membranes were prepared by depositing a thin TPIM layer on top of crosslinked P84® poly(imide) substrates by means of spin coating. With the modification using trimesoyl chloride, the resultant multilayer membranes have a rejection reaching 90% to Remazol brilliant blue R with a pure ethanol permeance of 3.4 $l\,m^{-2}bar^{-1}h^{-1}$. It also has an impressively high acetone permeance of 12.42 $l\,m^{-2}bar^{-1}h^{-1}$ with a rejection of 97% to rose bengal (57).

3.13.4 *Separation of 1,3-Propanediol*

The synthesis of an allylcyclohexylamine-functionalized siloxane has been reported (58). Allylcyclohexylamine is shown in Figure 3.31.

The membrane structure is based on an allylcyclohexylamine-functionalized poly(hydromethylsiloxane). The fabrication can be done by two methods. In one method, it is crosslinked with a high molecular weight hydroxyl-terminated poly(hydromethylsiloxane). The other method entails the formation of a microphase separated

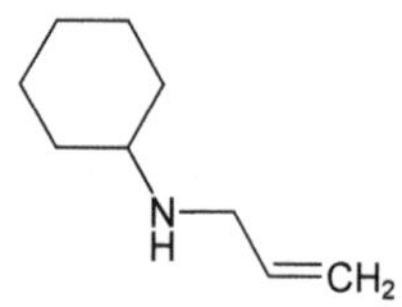

Figure 3.31 Allylcyclohexylamine.

semi-interpenetrating polymeric network blend with a high molecular weight styrene butyl acrylate copolymer.

Its phase-separated blend with a styrene butyl acrylate copolymer can be used for the pervaporative enrichment of 1,3-propanediol from dilute aqueous solutions (58).

The phase-separated blend allowed for the recovery of mechanical strength lost due to functionalization without loss in separation performance. Separation factors of 9 to 15 were achieved with functionalization levels of 50% to 90%, while the 1,3-propanediol flux was 5.5 $g\,m^{-2}h^{-1}$ to 5.8 $g\,m^{-2}h^{-1}$.

The separation efficiency increased with functionalization and decreased with increasing temperature and feed concentration. A solution diffusion model was used to compute the overall mass transfer coefficients, concentration polarization and intrinsic material mass transport properties. The overall mass transfer coefficient for 1,3-propanediol was between 1.0×10^{-7} to 1.4×10^{-7} $m\,s^{-1}$, while the boundary layer mass transfer coefficient ranged from 5×10^{-7} $m\,s^{-1}$ to 18×10^{-7} $m\,s^{-1}$, indicating the dominance of the membrane on the transport resistance (58).

3.13.5 *Separation of Carbon Dioxide*

One of the greatest challenges is to find a thermodynamically efficient process for separating and capturing CO_2 (59,60). It has been commonly accepted that global warming is related directly to the increased levels of CO_2 (61–63)

The CO_2 is mainly emitted as a result of the combustion of fossil fuels, certain chemical plant processes, and production of synthesis gas (64).

3.13.5.1 *Tetrazole-Functionalized Polymer Membranes*

A CO_2-philic tetrazole group functionalized polymer nanosieve membrane was reported for CO_2 capture applications (65). These membrane materials were prepared by a [2+3] cycloaddition modification of a polymer of intrinsic microporosity containing an aromatic nitrile group with an azide compound. The tetrazole group functionalized polymer nanosieve membranes showed enhanced CO_2-philic separation selectivities due to interactions between CO_2 and the tetrazole compared to a membrane from a polymer of intrinsic microporosity.

Membranes from polymers of intrinsic microporosity have shown exceptional properties, e.g., extremely high gas permeability, for the separation of commercially important gas pairs, including O_2/N_2 and CO_2/CH_4. The exceptionally high permeability of gases arises from the rigid but contorted molecular structures of polymers of intrinsic microporosity, frustrating packing and creating free volume, coupled with chemical functionality giving strong intermolecular interactions (66,67).

Membranes from polymers of intrinsic microporosity, however, have much lower selectivities for commercially important gas pairs, such as O_2/N_2 and CO_2/CH_4, although their gas permeabilities are significantly higher than those of commercial polymeric membranes from glassy polymers such as poly(imide)s (PIs) and PEIs (68).

A polymer containing nitrile groups was synthesized from the monomers 3,3,3′,3′-tetramethyl-1,1′-spirobisindane-5,5′,6,6′-tetrol and 2,3,5,6-tetrafluoroterephthalonitrile, as shown in Figure 3.32, has been synthesized following a procedure reported in the literature (69).

An efficient dibenzodioxane-forming reaction, i.e., aromatic nucleophilic substitution, between the aromatic tetrol monomer with the fluorine-containing compound resulted in a soluble polymer with intrinsic microporosity in a high yield. The polymer is freely soluble in methylene chloride, THF, and chloroform. The polymer was purified by repeated precipitation from THF solution into methanol and when collected by filtration gave a fluorescent yellow free-flowing powder (68).

The UV-crosslinked polymer membranes are especially useful in gas separation processes in air purification, petrochemical, refinery,

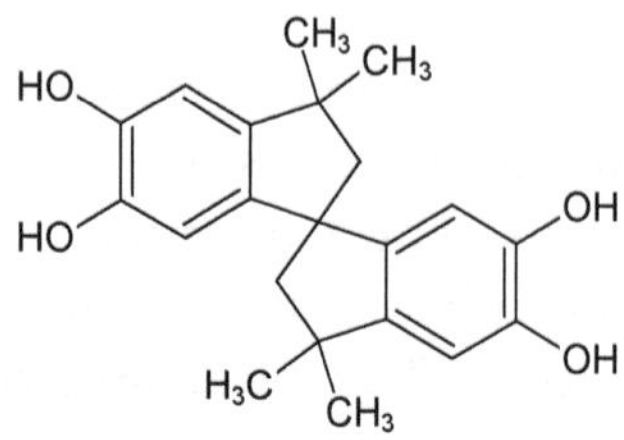

3,3,3′,3′-Tetramethyl-1,1′-spirobisindane-5,5′,6,6′-tetrol

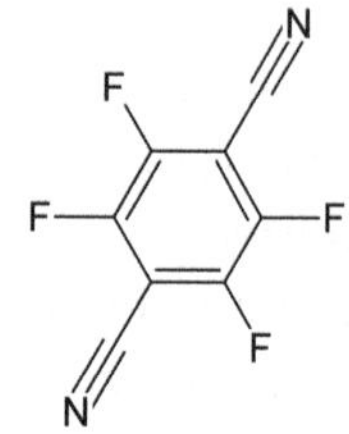

2,3,5,6-Tetrafluoroterephthalonitrile

Figure 3.32 Monomers (68).

and natural gas industries. Examples of such separations include the separation of volatile organic compounds, such as toluene, xylene, and acetone, from an atmospheric gas, such as nitrogen or oxygen and nitrogen recovery from air (68).

3.13.5.2 *Separation of Carbon Dioxide and Methane*

The performance of CO_2 capture from methane was studied under various operating conditions, including high pressure to investigate the effects of a heavy hydrocarbon (pentane) on the CO_2 capture and separation performance of membranes (70).

Based on these findings, the permeance and selectivity of CO_2 and of CH_4 under wet conditions (with heavy hydrocarbon) decrease compared to dry conditions (without heavy hydrocarbon). Therefore, the presence of pentane as a heavy hydrocarbon in a natural gas stream decreases the performance of CO_2 capture when using membrane processes (70).

A significant enhancement in the separation performance of supported poly(aniline) and poly(pyrrole) membranes after chemical modification via grafting and solvation with a non-volatile liquid, poly(ethylene glycol), has been demonstrated (59).

3.13.5.3 *Separation of Carbon Dioxide and Nitrogen*

A ionic liquid composite from 1-butyl-3-methylimidazolium tetrafluoroborate, c.f. Figure 3.33, and CuO was prepared for CO_2 separation membranes (60).

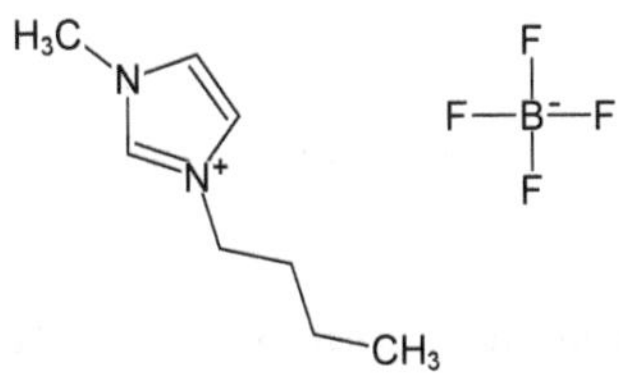

Figure 3.33 1-Butyl-3-methylimidazolium tetrafluoroborate.

The ideal CO_2/N_2 selectivity in the case of the composite material was 21.0, with a CO_2 permeance of 52.4 *GPU*, while the neat 1-butyl-3-methylimidazolium tetrafluoroborate membrane showed a selectivity of 5.0 and a CO_2 permeance of 17.0 *GPU* (60). Here, *GPU* is the *gas permeation unit* (71).

The enhanced separation performance was attributed to the synergistic effect of well-dispersed CuO nanoparticles, resulting in enhanced CO_2 solubility due to the increased surface of the oxide particle layer and the abundant free ions in the composite (60).

3.13.5.4 *Covalent Immobilization of Enzymes*

Reactive polymer materials were developed which are suitable for use as supports for enzyme covalent immobilization (72). Here, bicomponent polymer membranes were developed using poly(acrylonitrile-*co*-vinyl acetate) (PAN-co-PVAc) mixed with PVA.

First, the blends were dissolved in dimethyl sulfoxide, until a homogeneous polymer solution was obtained. To prepare the membranes, these solutions were cast on a glass plate, followed by the immersion of this plate in a coagulation bath containing a 50%:50% by volume water:isopropyl alcohol mixture. In this way, membranes with OH functional groups were obtained. Before tyrosinase immobilization, the membranes were functionalized with glutardialdehyde, c.f. Figure 3.34. So, CHO binding sites were inserted and the membrane became reactive for the enzyme.

The modifications produced by these reactions were investigated by various analytical techniques. The occurrence of important

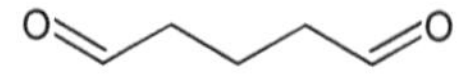

Figure 3.34 Glutardialdehyde.

changes in membrane features confirmed the success of the modification reactions. Furthermore, the activity of the bonded enzyme was determined by the pyrocatechol method (72).

Immobilization of 3-Hydroxybenzoate 6-Hydroxylase. 3-Hydroxybenzoate 6-hydroxylase is an enzyme that catalyzes the regiospecific *p*-hydroxylation of aromatic compounds (73).

This enzyme could be successfully immobilized onto electrospun poly(caprolactone) (PCL) ultrafine fibers of 424±99 *nm* in diameter (73). The fibers were fabricated from 13% PCL with a molecular weight of around 80 *k Dalton*, dissolved in a mixed solvent of formic acid (25% v/v) and acetic acid (75% v/v) at an applied voltage of 16 *kV* and a fiber collection distance of 12.5 *cm*.

Before being immobilized with 3-hydroxybenzoate 6-hydroxylase, the surface of electrospun PCL fibers was functionalized by the reaction with ethylene diamine activated with 5% v/v glutaraldehyde using aluminum sulfate as a catalyst.

The effects of the immobilization process on the pH tolerance and thermal stability of 3-hydroxybenzoate 6-hydroxylase were investigated. The results of these experiments indicated that 3-hydroxybenzoate 6-hydroxylase immobilized onto PCL fibers could tolerate the changes in temperature and pH better than the free enzyme. Therefore, the 3-hydroxybenzoate 6-hydroxylase immobilized PCL fibers are potentially useful as a heterogeneous catalyst under the conditions in which the free 3-hydroxybenzoate 6-hydroxylase enzyme could not endure (73).

3.13.5.5 Dye-Functionalized Polymer Membranes

Artificial photosynthesis could be a cost-effective and sustainable means of converting sunlight energy into usable energy (74). Most artificial photosynthetic systems mimic the properties found in nature of light absorption, electronic charge separation, and electronic charge collection, where ultimately electrons make and break chemical bonds. However, nature also transduces photon energy into proton gradients whose electric potential also drives chemical-bond formation.

The mimicking of nature in function has been reported, by using light to drive endergonic proton transfer, ultimately converting photon energy into ionic current (74). The applicability and practicality of this material as a membrane in solar fuel devices and as a separate ionic photoelectrochemical device have also been discussed.

A conical nanopore etched in a poly(ethylene terephthalate) plastic sheet was used. The pores were formed via swift heavy ion bombardment followed by alkaline chemical etching, which resulted in a pore lined with fixed anionic functional groups. Then, using peptide coupling chemistries, the pore was asymmetrically functionalized with photoacids to generate a region containing fixed cationic dyes.

This ordered distribution of interfacial charge was intended to mimic a solid-state semiconductor *pn*-junction and for use in the separation of photogenerated charges. Under visible-light illumination and a small reverse bias, excitation of the photoacid molecules resulted in an ionic photocurrent.

The development of an artificial light-driven ion pump to other polymeric materials has begun which would generate ionic power through sunlight absorption and incorporate these materials as ion exchange membranes in solar fuel devices (74).

3.13.5.6 Ultrafiltration Membranes

Ultrafiltration membranes were prepared by blending brominated poly(phenylene oxide) (PPO) and its quaternary phosphonium derivative (TPPOQP-Br) as additive using a phase inversion method (75).

X-ray photoelectron spectroscopy (XPS) studies indicated that the brominated PPO/TPPOQP-Br composite membranes exhibited an increase in the concentration of TPPOQP-Br from the top surface to the bottom surface. In contrast, the composite membranes prepared from BPPO and its quaternary ammonium derivative (TPPOQA-Br) showed an opposite concentration gradient than that of

TPPOQA-Br. This was attributed to the difference in wettability and hydration rate between TPPOQP-Br and TPPOQA-Br, leading to a different membrane microstructure and chemical composition distributions.

The brominated PPO membrane shows a water flux of 215 $l\,m^{-2}h^{-1}$ at 100 kPa and its molecular weight cut-off of PEG is 93.8 kDa. The corresponding values of the optimal brominated PPO/TPPOQP-Br membrane are 873 $l\,m^{-2}h^{-1}$ and 111.3 kDa, both of which are better than those of brominated PPO/TPPOQA-Br with similar additive loading.

Therefore, the addition of TPPOQP-Br significantly enhances the water permeability while maintaining the excellent rejection properties in the resultant ultrafiltration membranes.

Thus, the addition of a hydrophobic and charged polymer with slow hydration property is an effective strategy for improving flux and anti-biofouling properties of ultrafiltration membranes (75).

3.13.5.7 *Low-Humidity Proton Conducting Membranes*

Polyelectrolyte membranes with a proton conductivity at moderate levels of relative humidity and temperature are essential for the development of polyelectrolyte membrane fuel cells (76).

Monomethoxy oligo ethylene glycol methacrylate derived polymer brush-functionalized silica nanoparticles were developed as humidifying nanoadditives for the fabrication of Nafion nanocomposite membranes (76). These materials exhibit improved proton conductivities at moderate levels of relative humidity and temperature.

Polymer brush-functionalized SiO_2 nanoparticles (SiO_2-polymer brush), fabricated via surface-initiated atom transfer radical polymerization (ATRP), are dispersed in the Nafion resin solution, and nanocomposite membranes (Nafion/SiO_2-polymer brush) are fabricated via solution-casting. For comparative studies, composite membranes of Nafion were also prepared with bare SiO_2 nanoparticles.

The water uptake capacity was determined by submerging the membranes in deionized water for 8 h followed by drying at 50°C under vacuum for 8 h. The difference in weight between wet and dry membranes was measured.

Spectroscopic measurements confirmed the presence of polymer brushes in the final membranes and demonstrated an increased water uptake in membranes with polymer brush-functionalized nanocomposite membranes.

Electrochemical impedance analysis revealed that 1% of functionalized SiO_2 nanoparticles is sufficient to achieve Nafion nanocomposite membranes with superior proton conductivities at ambient and moderately high temperatures over the entire range of relative humidity (76).

3.13.5.8 Sulfonated Compounds

Sulfonated Propylsilane Graphene Oxide. Sulfonated poly(imide)/sulfonated propylsilane graphene oxide was assessed to be a promising candidate for polymer electrolyte membranes (77).

The incorporation of a multifunctionalized ($-SO_3H$ and $-COOH$) sulfonated propylsilane graphene oxide in a sulfonated poly(imide) matrix improved the proton conductivity and thermal, mechanical, and chemical stabilities along with bound water content responsible for slow dehydration of the membrane matrix.

The sulfonated propylsilane graphene oxide/sulfonated poly(imide) composite polymer electrolyte membrane was designed to promote internal self-humidification, responsible for water-retention properties, and to promote proton conduction, due to the presence of different acidic functional groups. Strong hydrogen bonding between multifunctional groups thus led to the presence of interconnected hydrophobic graphene sheets and organic polymer chains, which provides hydrophobic-hydrophilic phase separation and suitable architecture of proton-conducting channels. In single-cell direct methanol fuel cell tests, sulfonated poly(imide)/sulfonated propylsilane graphene oxide-8 exhibited 75.06 $mW\,cm^{-2}$ maximum power density (in comparison with commercial Nafion 117 membrane, 62.40 $mW\,cm^{-2}$) under 2 M methanol fuel at 70°C (77).

Zwitterionic Polymer-functionalized Graphene Oxide. Hybrid membranes of sulfonated poly(imide) (SPI) and zwitterionic polymer-functionalized graphene oxide (ZGO) were fabricated via a solution-casting method for a vanadium redox flow battery (78). The successful preparation of ZGO fillers and sulfonated poly(imide)/ZGO hybrid membranes could be demonstrated by FTIR, XPS and SEM, indicating that the ZGO fillers were homogeneously dispersed into the sulfonated poly(imide) matrix.

Through controlling the interfacial interaction between sulfonated poly(imide) matrix and ZGO fillers, the physicochemical properties, e.g., vanadium ion barrier and proton transport pathway, of the hybrid membranes are tuned via the zwitterionic acid-base interaction in the hybrid membrane. This shows a high ion selectivity and good stability with the incorporated ZGO fillers.

Sulfonated poly(imide)/ZGO-4 hybrid membrane is proven to have higher cell efficiencies (CE: 92–98%, EE: 65–79%) than commercial Nafion 117 membrane (CE: 89–94%, EE: 59–70%) for vanadium redox flow battery application at 30–80 $mA\,cm^{-2}$. The assembled vanadium redox flow battery with sulfonated poly(imide)/ZGO-4 membrane presents a stable cycling charge/discharge performance for over 280 cycles, which demonstrates its excellent chemical stability under strong acidic and oxidizing conditions. Sulfonated poly(imide)/ZGO hybrid membranes have brilliant prospects (78).

3.13.6 Adsorption of Methylene Blue

A diversity of nanofibers can be fabricated with electrospinning technology. This technology promotes applications in various fields, including wastewater treatment (79).

Biodegradable poly(butylene succinate-*co*-terephthalate) nanofibrous membranes were fabricated by electrospinning and functionalized them with a β-cyclodextrin polymer, c.f. Figure 3.35, using *in-situ* polymerization of β-cyclodextrin polymer on the surface of poly(butylene succinate-*co*-terephthalate) nanofibrous membranes.

A much higher efficiency was found for removing the methylene blue dye on the resultant poly(butylene succinate-*co*-terephthalate)/β-cyclodextrin polymer nanofibrous membranes than that on pure β-cyclodextrin polymer, or a poly(butylene succinate-*co*-terephthalate) nanofibrous membrane. Methylene blue is shown in Figure 3.36.

The adsorption performance of the composite membranes to methylene blue molecules was influenced greatly by the amount of β-cyclodextrin polymer that dominated the surface area and surface morphologies of the prepared membranes. The adsorption

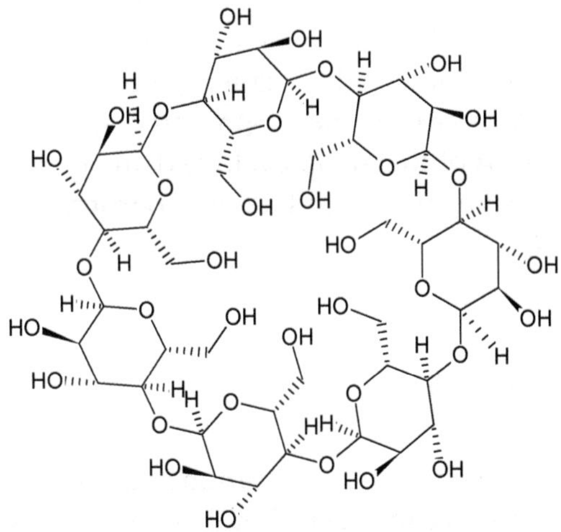

Figure 3.35 β-Cyclodextrin.

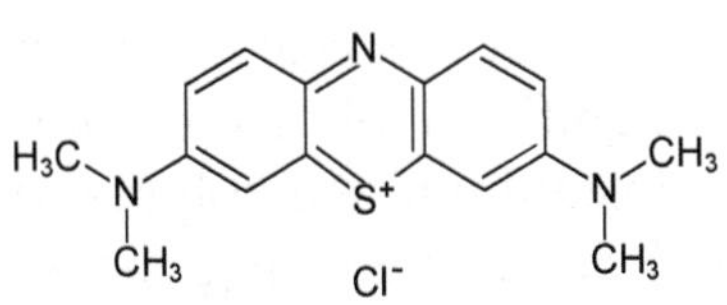

Figure 3.36 Methylene blue.

kinetics of poly(butylene succinate-*co*-terephthalate)/β-cyclodextrin polymer nanofibrous membrane fitted well with the pseudo-second-order model, and the Langmuir isotherm model well described the relationship between the adsorption capacity and initial methylene blue solution concentration, exhibiting the maximum adsorption capacity of 90.9 $mg\,g^{-1}$, which was much higher than the adsorption capacity of the other adsorbents. Considering the excellent adsorption performance of the so-prepared material, poly(butylene succinate-*co*-terephthalate)/β-cyclodextrin polymer nanofibrous membranes are promising candidates in environmental purification applications (79).

3.13.7 *Water Permeability*

The prediction and control of penetration in a polymeric membrane is of critical importance in green chemistry and energy technology, including gas separation, water purification, and desalination (80).

Molecular simulations of water transport through a polymeric membrane were performed to clarify the key factors that dominate the water permeation. Also, the effects of additives and chemical interaction (solubility) on water inhibition were investigated. It was found that certain additives can reduce water permeability into the membrane. Upon the incorporation of the additive, strength of coordination of water molecules near the membrane surface increases. Thus, the penetration frequency of water molecules into the membrane decreases.

It has been suggested that the local environment near the membrane surface plays a significant role in controlling water permeability. To gain a deeper insight into the polymer design, the chemical interaction (solubility) parameter change between polymer chains and additives was discussed (80). The use of a repulsive chemical species of a polymer chain for additives can lead to a higher water inhibition.

However, it has been suggested that the water permeability can be controlled to some extent without synthesizing a novel functional polymer (80).

3.13.8 *Porous Molecularly Imprinted Polymers*

Magnetic porous molecularly imprinted polymers (MPMIPs) for rapid and efficient selective recognition of chlorogenic acid could be effectively prepared using surface precipitation polymerization using chlorogenic acid as template, 4-vinylpyridine as a functional monomer, and a mesoporous SiO_2 layer as sacrificial support (81). These compounds are shown in Figure 3.37.

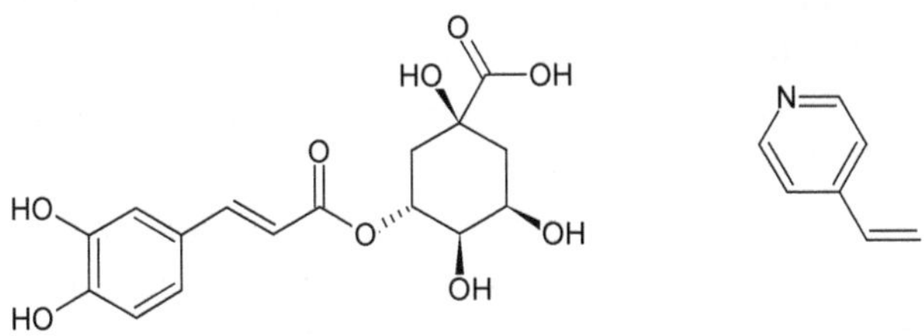

Chlorogenic acid 4-Vinylpyridine

Figure 3.37 Surface precipitation polymerization materials (81).

A computational simulation by evaluation of the electronic binding energy was used to optimize the stoichiometric ratio between chlorogenic acid and 4-vinylpyridine (1:5), which reduced the duration of the laboratory experiments. The porous molecularly imprinted polymer shell and the elimination of solid molecularly imprinted polymers by magnet gave magnetic porous molecularly imprinted polymers high binding capacity of 42.22 $mg\,g^{-1}$ and a fast kinetic binding of 35 *min*. The adsorption behavior between chlorogenic acid and magnetic porous molecularly imprinted polymers followed the Langmuir equation and a pseudo-first-order reaction kinetics.

Furthermore, the obtained magnetic porous molecularly imprinted polymers as solid-phase adsorbents coupled with high-performance liquid chromatography were employed for the selective extraction and determination of chlorogenic acid ($2.93 \pm 0.11\ mg\,g^{-1}$) in Duzhong brick tea samples. Recoveries from 91.8% to 104.2% and a detection limit of 0.8 $\mu g\,ml^{-1}$ were obtained. The linear range of 2.0–150.0 $\mu g\,ml^{-1}$ was wide, with R2 > 0.999. Overall, this study provided an efficient approach for fabrication of well-constructed magnetic porous molecularly imprinted polymers for fast and selective recognition and determination of chlorogenic acid from complex samples (81).

3.13.9 *Membranes for Olefin/Paraffin Separation*

The separation of gases from a multicomponent mixture of gases is necessary in a large number of industries (82). Such separations can be performed by processes such as cryogenics, pressure swing adsorption, and membrane separations. In certain types of gas separation, membrane separation has been found to be economically more viable than other processes.

In a pressure-driven gas membrane separation process, one side of the gas separation membrane is contacted with a multicomponent gas mixture. Certain gases of the mixture can permeate through the membrane faster than the other gases. So, gas separation membranes allow some gases to permeate through them while serving as a relative barrier to other gases. The relative gas permeation rate through the membrane is a property of the membrane material composition and its morphology.

Much of the work in this field has been directed to the development of membranes that optimize the separation factor and total flux of a given system. It has been shown that aromatic PIs containing the residue of alkylated aromatic diamines are useful in separating a variety of gases (83). Moreover, it has been reported that other poly(imide)s, poly(carbonate)s, PUs, polysulfones and poly(phenyleneoxide)s are useful for like purposes.

In the petrochemical industry, one of the most important processes is the separation of olefin and paraffin gases (82). Olefin gases, particularly ethylene and propylene, are important chemical feedstocks. Various petrochemical streams contain olefins and other saturated hydrocarbons. These streams typically originate from a catalytic cracking unit. The separation of olefin and paraffin components can be done using low temperature distillation. Distillation columns are normally around 300 feet tall and contain over 200 trays. This is extremely expensive and energy intensive due to the similar volatilities of the components. Thus, membrane separation methods have been considered as an attractive alternative (82).

A polymeric composite has been described that may be used for fluid separation membranes (82). The fluid separation membrane may be formed from the reaction product of a tetraacid compound and a diamine. Tetraacid compounds are shown in Figure 3.38 and tetraamino compounds are shown in Figures 3.39 and 3.40. In addition, a thiolene structure is shown in Figure 3.41.

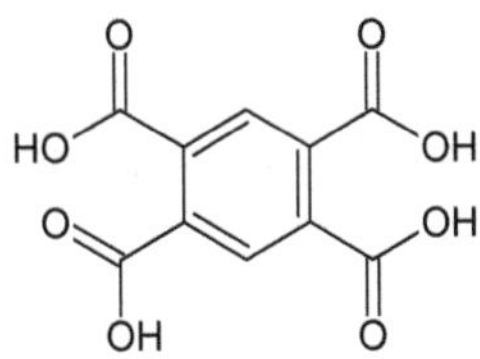

1,2,4,5-Benzene tetracarboxylic acid

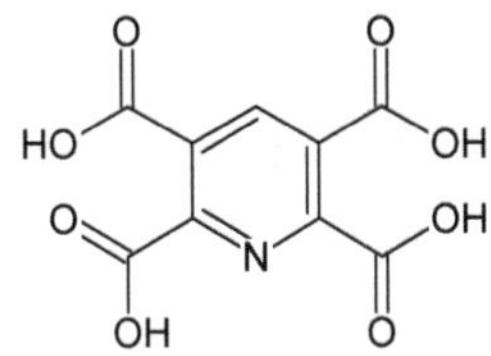

Pyridine tetraacid

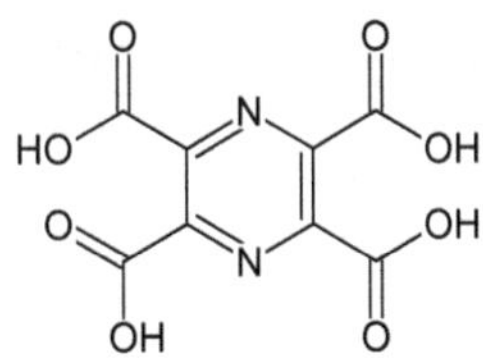

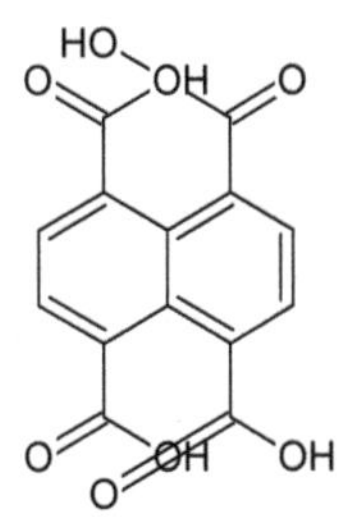

Pyrazine tetraacid 1,2,5,6-Naphthalene tetracarboxylic acid

Figure 3.38 Tetraacid compounds (82).

The initial resulting product is a poly(amide) (PA). The PA may be thermally or chemically cyclized to form a PI. The polymer matrix of the fluid separation membrane may also include a dithiolene.

PAs may be formed by the condensation of a tetraamine, a tetraacid, and a diamine. Poly(imide)s and poly (pyrrolone-imides) may be formed by the cyclization of a polymer precursor. A polymeric composite may include a dithiolene or a mixture of dithiolenes. A polymer matrix incorporating dithiolenes may exhibit an olefin/paraffin solubility selectivity. A solubility selectivity may be between about 1.1 and about 2.0 (82).

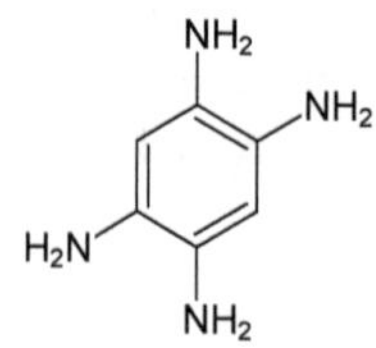

1,2,4,5-Tetraminobenzene

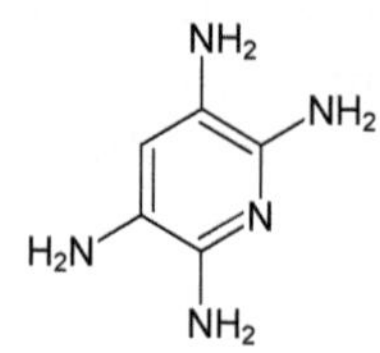

1,2,4,5-Tetraminopyridine

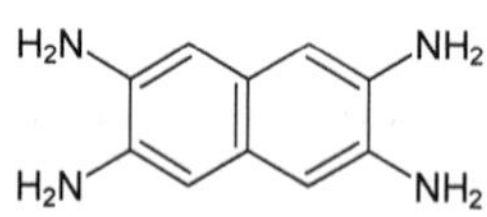

2,3,6,7-Tetraaminonaphthalene

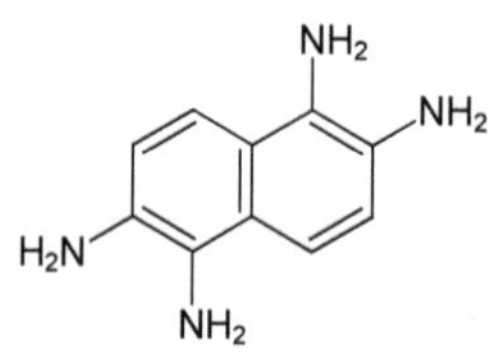

1,2,5,6-Tetraminonaphthalene

Figure 3.39 Tetraamino compounds (82).

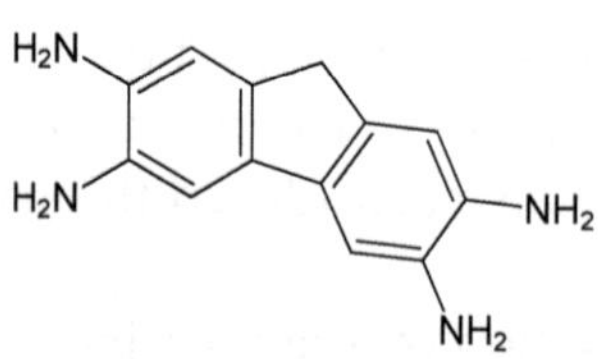

Tetraminofluorene

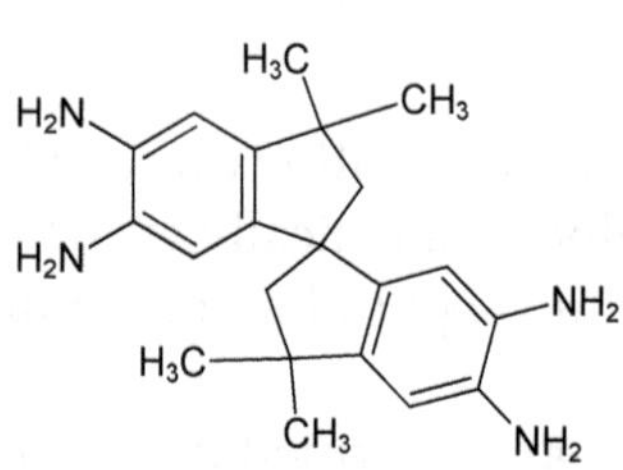

Tetramethyl-spiro-biindane

Figure 3.40 Tetraamino compounds (82).

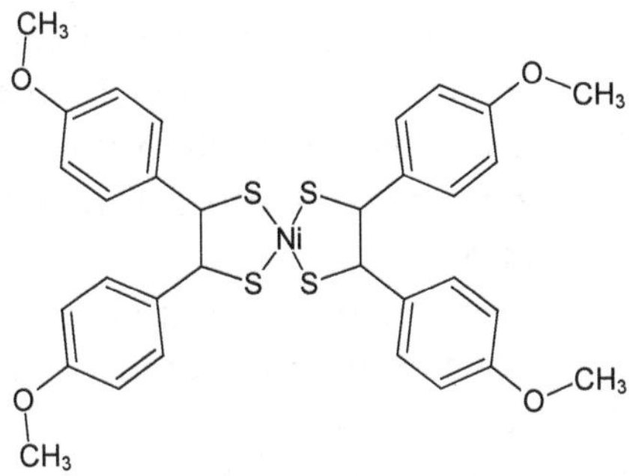

Figure 3.41 Thiolene (82).

3.13.10 *Thermally Responsive Ion-Permeable Membranes*

Thermally responsive separation membranes have been developed (84). The thermally responsive materials contain upper critical solution temperature (UCST) polymers that are covalently bound to a support substrate. The UCST polymers undergo a reversible conformational change above their UCST, which results in a reduction in the ionic permeability of such thermally responsive materials and membranes.

In energy storage applications, thermally responsive membranes can be used to mitigate a thermal runaway by suppressing the ion mobility at elevated temperature, thereby reducing the charge and discharge rates and the generation of heat. In other applications, they can be used to provide the function of internal cell-balancing by restricting the local ion flow.

Thin membranes that contain thermally responsive materials can be used as porous separation membranes for liquid electrolyte batteries, such as Li-ion batteries, Li-sulfur batteries and Li-air batteries (84).

During normal battery operation, the UCST polymers in the thermally responsive material are phase-segregated and form hydrophobic aggregates attached to the substrate. In this configuration, ion flux is readily permitted through open channels in the membrane. However, at higher temperatures, such as those that might occur during battery overheating, the polymer molecules become miscible with the electrolyte solvent, although the polymer chains remain covalently anchored to the substrate. In this configuration, the ion flux through the channels in the membrane is reduced. As a result, the ionic conduction through the separation membrane can be shut down at a temperature below that at which a thermal runaway occurs.

For such compositions, the main component is a poly(sulfobetaine) polymer that is covalently grafted onto a membrane and has an upper critical solution temperature (85,86).

Graphene oxide sheets are the structural component of the membrane. It has been proposed that at low temperatures the attractive dipolar interaction between zwitterions in the polymer causes the polymer to phase segregate and form hydrophobic aggregates attached to the membrane, thus permitting ion flux through open channels in the membrane. Above the UCST, when there is sufficient thermal energy to disrupt the attractive interaction, the polymer chains become independent and miscible with the electrolyte solvent, although one end of the polymer chain remains covalently anchored to the membrane.

The synthesis of graphene oxide functionalized with poly(3-(*N*-2-methacryloyloxyethyl-*N*,*N*-dimethyl)ammonatobutane-sulfonate) (GO-PMABS) has been detailed (84). The monomer is shown in Figure 3.42.

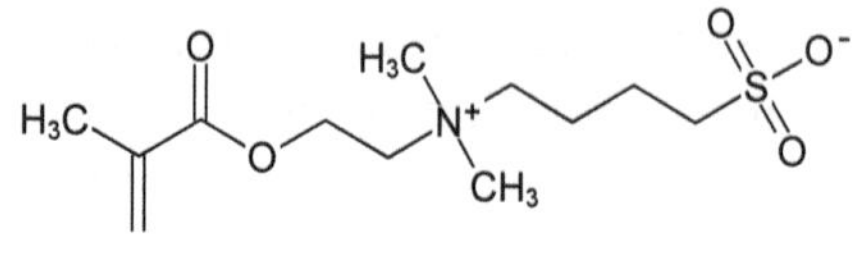

Figure 3.42 3-(*N*-2-Methacryloyloxyethyl-*N*,*N*-dimethyl)ammonatobutane-sulfonate.

Preparation 3–6: First, 10 *g* 2-bromo-2-methylpropanoate-4-methylphenethyl functionalized graphene oxide and 15 *g* 3-(*N*-2-methacryloyloxyethyl-*N*,*N*-dimethyl)ammonatobutane-sulfonate were mixed with 40 *ml* distilled deionized water and 160 *ml* MeOH at room temperature. Then the mixture was deoxygenated by bubbling dry argon through it while stirring for 30 min. While under Ar, the catalyst system consisting of 420 *mg* 2,2′-dipyridyl, c.f. Figure 3.43, 106 *mg* CuCl, and 15 *mg* $CuCl_2$ was added to initiate the polymerization reaction. The mixture continued to be stirred under Ar for another 30 *min*. After 72 *h*, the polymerization was terminated by exposing the mixture to air.

The product was filtered, washed with hot water several times to remove polymer that was not covalently bound, and dried at 60°C for 24 *h*. The polymer content of the final product estimated from TG was around 70%.

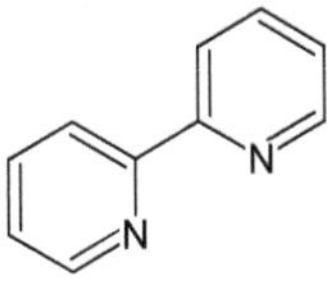

Figure 3.43 2,2′-Dipyridyl.

The UCST behavior of poly(3-(*N*-2-methacryloyloxyethyl-*N*,*N*-dimethyl)ammonatobutane-sulfonate) in water could be observed visually. At 20°C a 10% polymer mixture in water was cloudy, but at 50°C, it became clear (84).

In summary, it could be shown that a membrane that is modified with a UCST polymer can function to restrict the ion flow reversibly at elevated temperatures (84). This was demonstrated using a poly(3-(*N*-2-methacryloyloxyethyl-*N*,*N*-dimethyl)ammonatobutane-sulfonate) modified graphene oxide, and applied to an assembly that resembles a Li-ion battery. Since the upper critical temperature depends on the nature of the polymer and the solvent, such as molecular weight and density and nature of the zwitterions, there are many avenues available to tune the transition temperature and magnitude of change of ionic permeability. For example, the transition temperature would be higher for a larger molecular weight polymer with a higher density of zwitterions (84).

3.13.11 *Janus Graphene Oxide/Chitosan Hybrid Membranes*

A robust and simple method has been reported for the preparation of polymer brush-functionalized Janus graphene oxide/chitosan hybrid membranes via the combination of interface self-assembly of graphene oxide and chitosan, with subsequent self-initiated photografting and photopolymerization from both sides of the graphene oxide/chitosan composite membrane (87).

Self-initiated photografting is a facile polymerization method in which a surface-bound initiator is not required, and many vinyl monomers can be grafted from the photoactive site on the surface of GO/chitosan membrane by self-initiated photografting. Therefore, the compositions of Janus membrane surfaces are alternative in a wide range and the surface properties of GO/chitosan can be tuned in a definite direction by selecting the appropriate vinyl monomers (87).

The composite membrane was obtained by interface self-assembly of graphene oxide and chitosan induced by electrostatic interaction, PS and poly(*N*,*N*-dimethylaminoethyl methacrylate) were then grafted from the photoactive sites of the upper surface and lower surface of the graphene oxide/chitosan membrane by self-initiated photografting, respectively (87). The monomer *N*,*N*-dimethylaminoethyl methacrylate is shown in Figure 3.44.

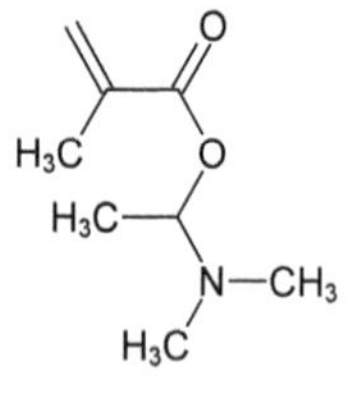

Figure 3.44 *N*,*N*-Dimethylaminoethyl methacrylate.

3.14 Rubber Formulations and Tire Materials

In order to reduce the rolling resistance and to improve the tread wear characteristics of tires, functionalized elastomers having a high rebound physical property, i.e., a low hysteresis, have been used for tire tread rubber compositions. However, in order to increase the wet skid resistance of a tire tread, rubbery polymers that have a relatively lower rebound physical property (higher hysteresis), which thereby undergo a greater energy loss, have sometimes been used for such tread rubber compositions.

To achieve such relatively inconsistent viscoelastic properties for the tire tread rubber compositions, blends of various types of synthetic and natural rubber can be utilized in tire treads.

Functionalized rubbery polymers made by living polymerization techniques are typically compounded with sulfur, accelerators, antidegradants, a filler such as carbon black, silica or starch, and other desired rubber chemicals. The compositions are then subsequently vulcanized or cured into the form of a tire or a power transmission belt.

It has been established that the physical properties of such cured rubbers depend upon the degree to which the filler is homogeneously dispersed throughout the rubber. This is in turn related to the level of affinity that the filler has for the particular rubbery polymer. This can be of practical importance in improving the physical characteristics of rubber articles, which are made utilizing such rubber compositions. For example, the rolling resistance and traction characteristics of tires can be improved by improving the affinity of carbon black and/or silica for the rubbery polymer utilized therein.

Therefore, it would be highly desirable to improve the affinity of a given rubbery polymer for fillers such as carbon black and silica (88,89).

The interaction between rubber and carbon black has been attributed to a combination of physical absorption (van der Waals force) and chemisorption between the oxygen-containing functional groups on the carbon black surface and the rubber (90,91)

3.14.1 *End-Group Functionalization of Rubber Polymers*

The end-group functionalization of rubbery living polymers has been presented to improve their affinity for fillers such as carbon black and/or silica (88). Such functionalized polymers can be beneficially used in manufacturing tires and other rubber products where improved polymer/filler interaction is desirable. In tire tread compounds this can result in lower polymer hysteresis, which in turn can provide a lower level of tire rolling resistance.

Examples of monomers are collected in Table 3.2 and some of the compounds used are shown in Figure 3.45.

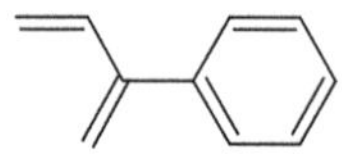

2-Phenyl-1,3-butadiene

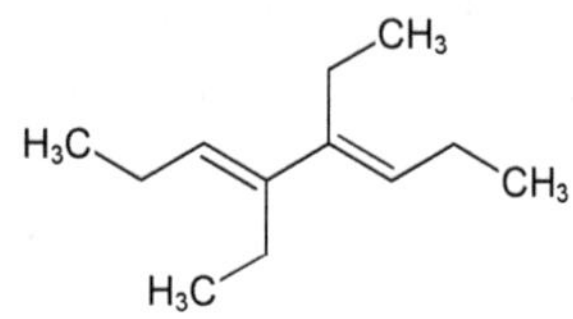

4,5-Diethyl-1,3-octadiene

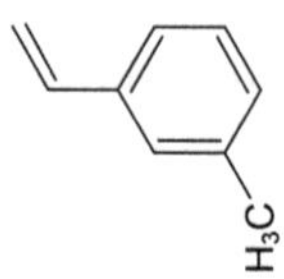

3-Methylstyrene

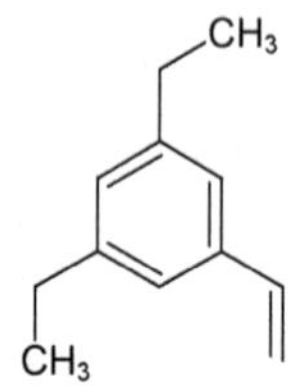

3,5-Diethylstyrene

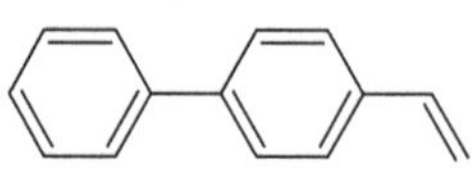

4-Phenylstyrene

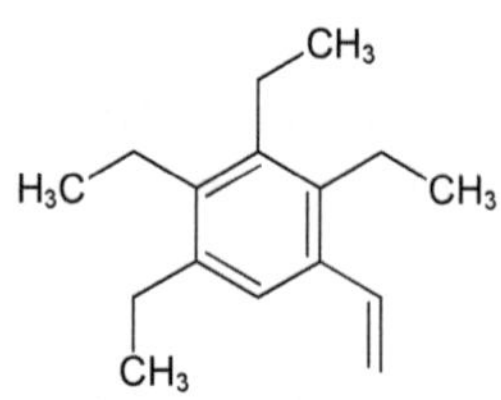

2,3,4,5-Tetraethylstyrene

Figure 3.45 Monomers (88).

Table 3.2 Monomers (88).

Conjugated diolefins	Styrenic monomers
1,3-Butadiene	Styrene
Isoprene	3-Methylstyrene
1,3-Pentadiene	3,5-Diethylstyrene
2,3-Dimethyl-1,3-butadiene	4-Propylstyrene
2-Methyl-1,3-pentadiene	2,4,6-Trimethylstyrene
2,3-Dimethyl-1,3-pentadiene	4-Dodecylstyrene
2-Phenyl-1,3-butadiene	4-Phenylstyrene
4,5-Diethyl-1,3-octadiene	2-Ethyl-4-benzylstyrene
	3,5-Diphenylstyrene
	2,3,4,5-Tetraethylstyrene
	α-Methylstyrene

The metal terminated living rubbery polymer can be functionalized by simply adding a stoichiometric amount of a multifunctional terminator to a solution of the rubbery polymer.

Examples of multifunctional terminators are collected in Table 3.3. Some of these compounds are shown in Figure 3.46

Table 3.3 Multifunctional terminators (88).

Compound
Ethoxysilatrane
1-Ethoxy-3,7,10-trimethylsilatrane
1-Isobutyl-3,7,10-trimethylsilatrane
1-Octyl-3,7,10-trimethylsilatrane
1-(3-Chloropropyl)-3,7,10-trimethylsilatrane
1,2-Bis(3,7,10-trimethylsilatrane)ethane
1,1′-(Decane-1,2-diyl)bis(3,7,10-trimethylsilatrane)
1,8-Bis(3,7,10-trimethylsilatrane)octane
1-Ethoxy-thiosilatrane
1-Ethoxy-2,8,9-triazasilatrane
1-(3-(Oxiran-2-ylmethoxy)propyl)-3,7,10-trimethylsilatrane

The synthesis of 1-ethoxy-3,7,10-trimethylsilatrane runs as follows (88):

Preparation 3–7: First, 253.0 *g* trisopropanolamine, 289 *g* tetraethyl silicate, and 3.7 *g* potassium hydroxide were mixed in a 1 *l* 3-neck round-bottom flask equipped with a distillation apparatus. The mixture was then

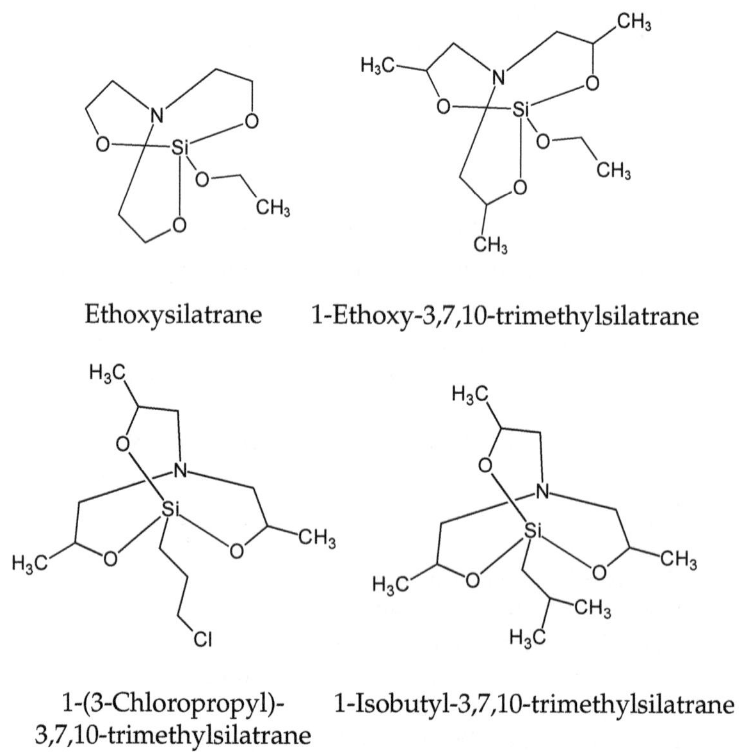

Figure 3.46 Multifunctional terminators.

heated to 85°C by a heating mantle, and ethanol produced from the reaction was removed under reduced pressure of 200 *mm Hg*. After 2 *h* of reaction the pressure was set to 100 *mm Hg*, and the mixture was heated to 120°C for an additional hour. A total amount of 230 *ml* ethanol was recovered from the distillation. The oily crude product was then distilled out under a pressure of 2 *mm Hg* at a temperature of 120°C. Finally, 305 *g* (88.2% yield) of a white crystalline solid, 1-ethoxy-3,7,10-trimethylsilatrane was obtained.

Furthermore, a bench-scale synthesis of a functionalized elastomer can be done as follows (88):

Preparation 3–8: The polymerization was done in eight-ounce bottles at 65°C in a water bath. A monomer premix of styrene and butadiene of 100 *g*, 15%, with a styrene/butadiene ratio of 21/79 was charged into an eight-ounce bottle with hexane as solvent, followed by addition of tetramethylethylene diamine modifier (TMEDA) and 0.10 *ml*, 1.6 $mol\ l^{-1}$ *n*-butyllithium as initiator. After a polymerization time of 0.5 *h*, additional styrene was added (1% of total monomer weight, about 10 units per polymer chain). The bottle was placed back into the 65°C water bath for an additional 1 *h*, and then the polymerization was terminated with functional terminators 1-ethoxy-3,7,10-trimethylsilatrane.

Tetramethylethylene diamine is shown in Figure 3.47.

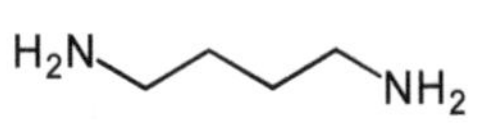

Figure 3.47 Tetramethylethylene diamine.

It was shown that the presence of a styrene sequence at the end of the living polymer prior to termination favors the formation of monochains terminally functionalized with tetraethyl orthosilicate rather than multiple chains coupled to a common tetraethyl orthosilicate terminator (88). Tetraethyl orthosilicate is shown in Figure 3.48

3.14.2 *Conjugated Diene-Based Polymers*

In the vehicle industry, the demand for the durability, stability and fuel economy of vehicles is continuously increasing, and much effort is directed toward satisfying this demand (92).

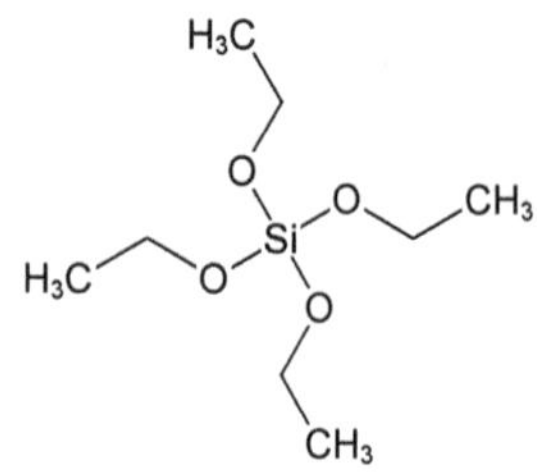

Figure 3.48 Tetraethyl orthosilicate.

In particular, many attempts have been made to enhance the properties of rubber as a material for vehicle tires, especially tire treads, which come in contact with roads. The rubber composition for a vehicle tire contains a conjugated diene-based polymer, such as poly(butadiene) or butadiene-styrene copolymer.

Thorough research is currently ongoing into the addition of various reinforcing agents to conjugated diene-based rubber compositions to increase the performance of vehicle tires. Specifically, as vehicles are required to exhibit stability, durability and fuel economy, rubber compositions having high processability and mechanical strength, including wear resistance, are being developed as material for vehicle tires, especially tire treads, which come in contact with roads.

In order to accomplish the above objectives, an end-modified conjugated diene-based polymer has been developed that is configured such that the end of a conjugated diene-based polymer is coupled with an aminosilane-based end modifier (92). For example, diethyl 3,3′-((3-(triethoxysilyl)propyl)azanediyl)dipropionate is shown in Figure 3.49.

The preparation of diethyl 3,3′-((3-(triethoxysilyl)propyl)azanediyl)dipropionate has been described as follows (92):

Preparation 3–9: First, 23.26 *mmol* of 3-aminopropyl-triethoxysilane was dissolved in 10 *ml* of ethanol in a 50 *ml* round-bottom flask, 46.53 *mmol* of ethyl acrylate was added, and the resulting mixture was stirred at 80°C for 24 *h* in a nitrogen atmosphere. After termination of the reaction, the solvent was removed under reduced pressure, followed by vacuum distillation at 80°C, yielding 22.36 *mmol* (yield 96.1%) of diethyl 3,3′-((3-(triethoxysilyl)-propyl)azanediyl)dipropionate.

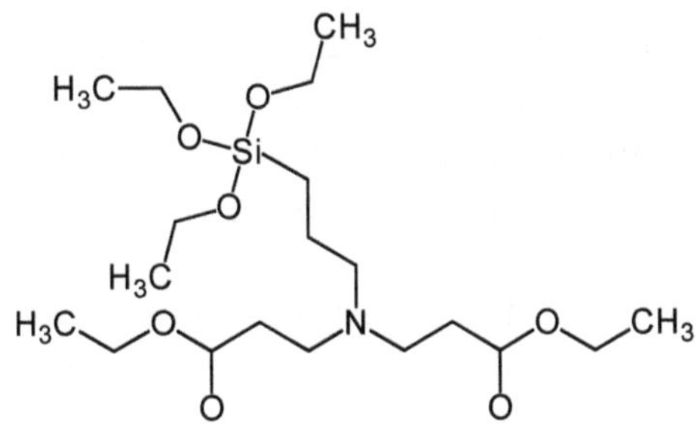

Figure 3.49 Diethyl 3,3′-((3-(triethoxysilyl)propyl)azanediyl)dipropionate.

Furthermore, the preparation of an end-modified styrene-butadiene polymer has been described as follows (92):

Preparation 3–10: First, 270 *g* of styrene, 710 *g* of 1,3-butadiene, 5 *kg* of *n*-hexane, and 1.1 *g* of 2,2-di(2-tetrahydrofuryl)propane as a polar additive were placed in a 20 *l* autoclave reactor, and the temperature inside the reactor was elevated to 40°C. When the temperature inside the reactor reached 40°C, 27 *g* (2.62% in hexane, 33% activation) of *n*-butyllithium was added to the reactor, followed by an adiabatic heating reaction. After about 30 *min*, 20 *g* of 1,3-butadiene was added so that the end of the material was capped with butadiene. After 5 *min*, 1.64 *g* of the modifier diethyl 3,3′-((3-(triethoxysilyl)propyl)azanediyl)dipropionate was added, and the reaction was carried out for 15 *min*. Thereafter, the polymerization was stopped using ethanol, and 33 *g* of a solution of a Wingstay K antioxidant dissolved at 30% in hexane was added. The resulting polymer was added to water warmed with steam, stirred to remove the solvent, and then roll-dried to remove the remaining solvent and water, yielding an end-modified conjugated diene-based polymer.

3.14.3 *Vulcanizates*

A method has been presented for the synthesis of a polymer having a terminal functionality (93). This method involves reacting a terminally active polymer with an α,β-ethylenically unsaturated compound that includes a group 2-13 element so as to provide a functionalized polymer.

The functionalized polymer can interact with various types of particulate filler, including carbon black and silica. Thus, the resulting polymer exhibits an enhanced interactivity with particulate fillers and can be used in the manufacture of vulcanizates.

Metal ester compounds can be used to synthesize functionalized polymers. The synthesis has been detailed in a study (93). Some of these compounds are collected in Table 3.4.

Table 3.4 Metal ester compounds (93).

Compound	Compound
Diisobutylaluminum acrylate	Ethylzinc acrylate
Diethyl boron acrylate	Diisobutylaluminum crotonate

For the synthesis of the functionalized polymers, first the monomers were polymerized and then functionalized with the compounds from Table 3.4.

3.14.4 Functionalized Polymer with Sulfide Linkage

Functionalized polymers have been used to reduce hysteresis loss and increase bound rubber. The functional group of the functionalized polymer is believed to reduce the number of polymer free ends. Also, the interaction between the functional group and the filler particles reduces the filler agglomeration, which thereby reduces hysteretic losses which are attributable to the disassociation of filler agglomerates, i.e., the Payne effect.

Methods for making polymers with a terminal functionality have been described. The terminal functionality includes a heteroatom and is connected to the polymer chain through a sulfide segment (94).

A carbanionic polymer chain that contains a polyene polymer can be synthesized with a terminal functionality separated from the polymer chain by a sulfide linkage (94).

The terminally functionalized polymer can be formed by reacting a carbanionic polyene-based polymer with an episulfide compound to yield a polymer with a terminal thiolate moiety, followed by the reaction with an epoxide compound that also includes other heteroatom-containing functional groups.

Depending on the intended end use, one or more of the polymer chains can include pendant aromatic groups, which can be used, for example, through incorporation of vinyl aromatic polymers, particularly the C_8 to C_{20} vinyl aromatic compounds, such as styrene, *α*-methylstyrene, *p*-methylstyrene, the vinyl toluenes, vinyl naphthalenes, and other related compounds (94). The issue of living anionic polymerization of functionalized styrene derivatives has been documented (95).

Solution polymerization typically uses an initiator, such as an alkali metal atom containing compound such as an organolithium compound, particularly alkyllithium compounds (94).

Examples of organolithium initiators are collected in Table 3.5.

Table 3.5 Organolithium initiators (94).

Compound	Compound
N-Lithio-hexamethylene imine	*n*-Butyllithium
Tributyltin lithium	Dimethylaminolithium
Diethylaminolithium	Dipropylaminolithium
Dibutylaminolithium	Diethylaminopropyllithium
2-Lithio-2-methyl-1,3-dithiane	2-Lithio-2-phenyl-1,3-dithiane
Trialkyl stanyl lithium compounds	

Other useful functional initiators include sulfur atom containing cyclic compounds (96). These are shown in Figure 3.50.

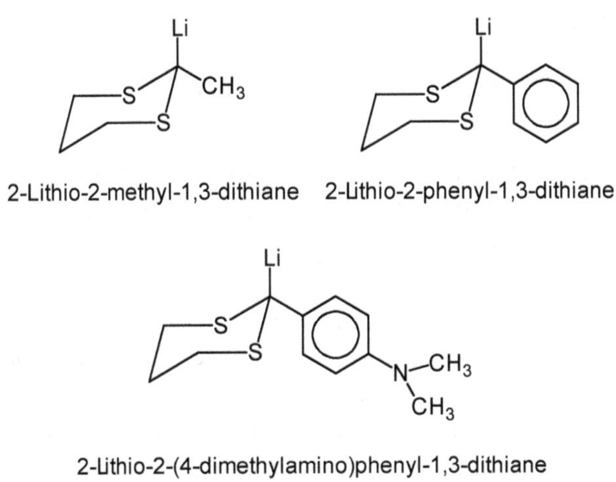

Figure 3.50 Sulfur atom containing cyclic functional initiators (96).

These initiators may be prepared by reacting an initiator precursor compound with an organolithium compound such as *n*-butyllithium. The synthesis of sulfur-functionalized initiators, specifically 2-lithio-2-methyl-1,3-dithiane and 2-lithio-2-phenyl-1,3-dithiane, from an initiator precursor and organolithium compound prior to polymerization can be done as follows (94):

Preparation 3–11: Commercially available solutions of 2-methyl-1,3-dithiane or 2-phenyl-1,3-dithiane are added to dried tetrahydrofuran, and cooled to approximately −78°C. A solution containing butyllithium and hexane is then added. The resulting solution is then stirred for approximately 3 *h* and allowed to stand overnight at a temperature of less than about 10°C. The resulting solutions may then be used to initiate the anionic polymerization. The preparation of the initiator may occur in an appropriate reaction vessel, including a polymerization reactor, prior to the addition of the solution of the monomers.

The synthesis of poly(styrene-*co*-butadiene) with 2-lithio-2-methyl-1,3-dithiane can be done as follows (94):

Preparation 3–12: To a 1.75 *l* N_2 purged reactor equipped with a stirrer was added 1.12 *kg* of hexane, 0.48 *kg* of 33% styrene in hexane, and 2.89 *kg* of 22.0% butadiene in hexane. The reactor was then heated to 24°C and 0.5 *ml* of 1.6 *M* of a cyclic oligomeric oxolanyl alkane modifier in hexane and 22.63 *ml* of 0.234 *M* 2-lithio-2-methyl-1,3-dithiane in tetrahydrofuran was charged to the reactor. The reactor jacket was then heated to 54°C. After 15 *min*, the batch temperature peaked at 76.5°C. After an additional 25 *min*, the cement was removed from the reactor, coagulated in isopropanol containing butylated hydroxy toluene, and drum dried to yield a polymer with the following properties: M_n=153 $kg\,mol^{-1}$, M_w=167 $kg\,mol^{-1}$, Tg=−44.4°C, 21.7% styrene, 1.3% block styrene, 32.1% vinyl, and 46.2% 1,4 butadiene incorporation.

Several other examples of preparation have been detailed (94). The prepared elastomeric compounds can optionally contain a silica coupling agent. Silica coupling agents are collected in Table 3.6 and shown in Figure 3.51.

Styrene-butadiene rubber (SBR) polymers were utilized to prepare a vulcanizable elastomeric compound with a combination of carbon black and silica as fillers.

A silica/carbon black compound with a SBR polymer prepared *in situ* with the initiator 2-lithio-2-(4-dimethyl-amino)phenyl-1,3-dithiane showed a 13% reduction in tan δ at 50°C, compared to a control compound containing the polymer prepared with an *n*-BuLi

Table 3.6 Silica coupling agents (94).

Compound
1-Mercaptomethyltriethoxysilane
2-Mercaptoethyltriethoxysilane
3-Mercaptopropyltriethoxysilane
3-Mercaptopropylmethyldiethoxysilane
2-Mercaptoethyltriproxysilane
18-Mercaptooctadecyldiethoxychlorosilane
Bis(3-triethoxysilyl-propyl)tetrasulfide

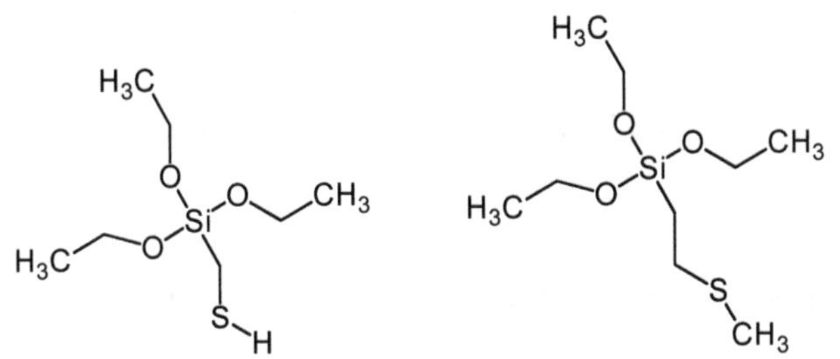

1-Mercaptomethyltriethoxysilane 2-Mercaptoethyltriethoxysilane

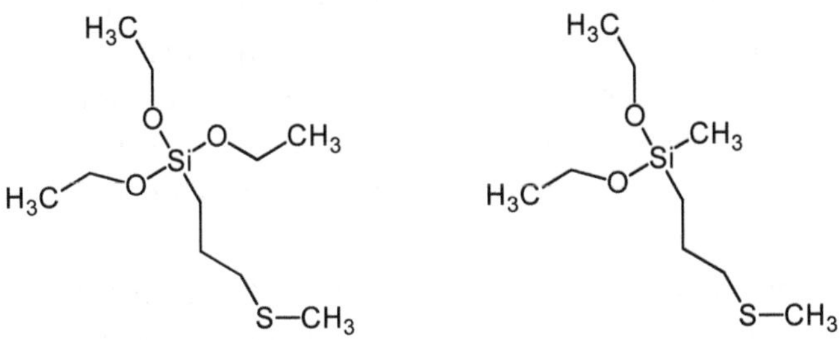

3-Mercaptopropyltriethoxysilane 3-Mercaptopropylmethyl-diethoxysilane

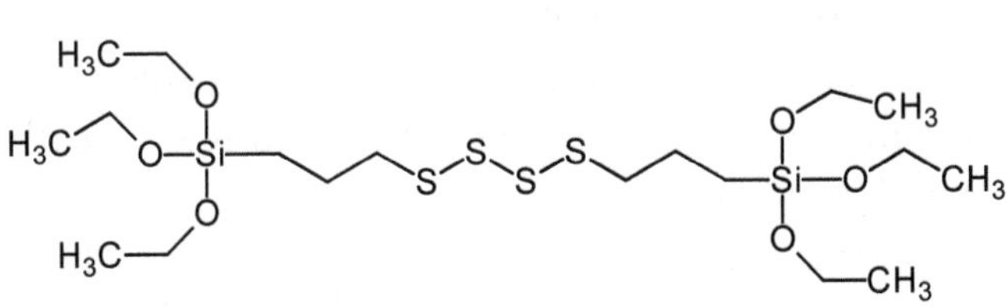

Bis(3-triethoxysilyl-propyl)tetrasulfide

Figure 3.51 Silica coupling agents.

initiator. A SBR 1,3-dimethyl-2-imidazolidinone containing silica/carbon black compound also showed a 24.4% reduction in tan δ at 50°C. compared to the *n*-Bu SBR 1,3-dimethyl-2-imidazolidinone containing silica/carbon black compound (94). 1,3-Dimethyl-2-imidazolidinone is shown in Figure 3.52.

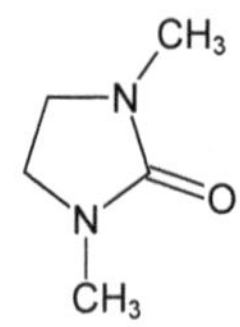

Figure 3.52 1,3-Dimethyl-2-imidazolidinone.

3.14.5 Polymer with a Hydrazine Functionality

Good traction and resistance to abrasion are primary considerations for tire treads (97). However, motor vehicle fuel efficiency concerns argue for a minimization in their rolling resistance, which correlates with a reduction in hysteresis and heat buildup during operation of the tire. These considerations are, to a great extent, competing and somewhat contradictory. Treads made from compositions designed to provide good road traction usually exhibit increased rolling resistance and vice versa.

A polymer has been described that includes a directly bonded hydrazone radical, which optionally can be located at a terminus of the polymer (97). This can be done by reacting a carbanionic living polymer with a hydrazone compound.

An exemplary elastomer for functionalization is SBR. A technique for incorporating hydrazine functionality is to react a carbanionic polymer with hydrazones.

Examples of monomeric hydrazine compounds for the preparation of hydrazones are collected in Table 3.7 and shown in Figure 3.53.

The condensation reaction to form the hydrazone can use a ketone or an aldehyde (97). Preferred hydrazones include those that result

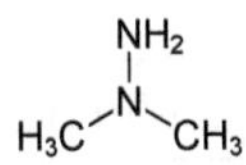

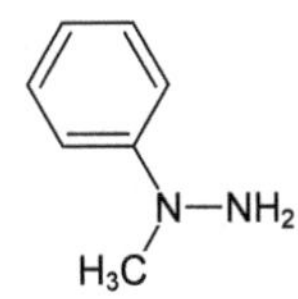

1,1-Dimethylhydrazine 1-Methyl-1-phenylhydrazine

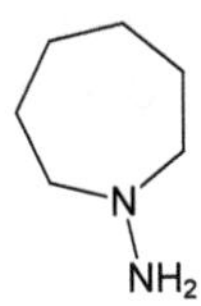

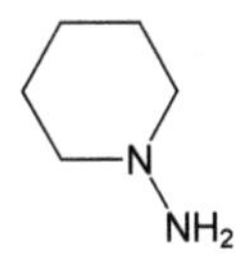

1-Aminohomopiperidine 1-Aminopiperidine

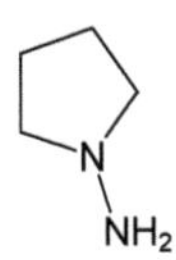

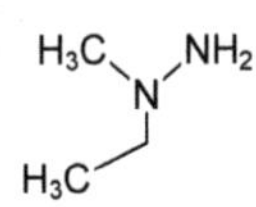

1-Aminopyrrolidine 1-Methyl-1-ethylhydrazine

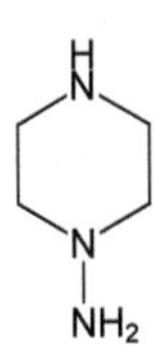

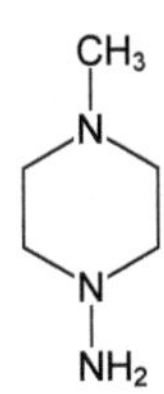

1-Aminopiperazine 1-Amino-4-methylpiperazine

Figure 3.53 Materials for hydrazone compounds (97).

Table 3.7 Materials for hydrazone compounds (97).

Hydrazine compounds	Amino-substituted heterocyclic compounds
1,1-Dimethylhydrazine	1-Aminopiperidine
1,1-Diethylhydrazine	1-Aminopiperazine
1-Methyl-1-ethylhydrazine	1-Amino-4-methylpiperazine
1-Methyl-1-phenylhydrazine	1-Aminopyrrolidine

from the condensation of an amino-substituted heterocyclic compound. The synthesis of cyclohexanecarboxaldehydepiperidinehydrazone from cyclohexanecarboxaldehyde and 1-aminopiperidine is shown in Figure 3.54.

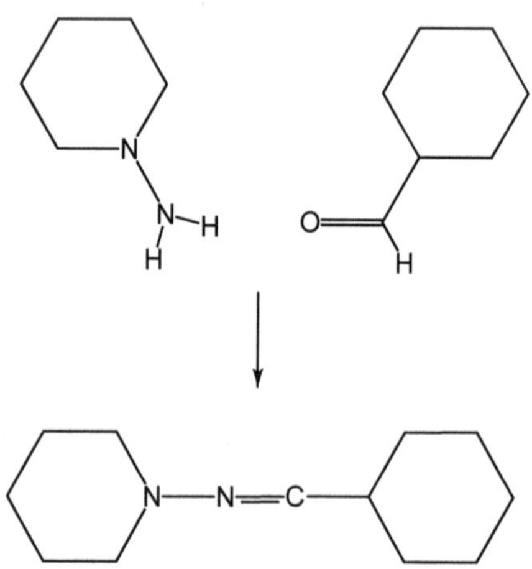

Figure 3.54 Synthesis of Cyclohexanecarboxaldehydepiperidinehydrazone (97).

Then, effecting the functionalization is to allow the hydrazone to react at the living terminal site of the polymer. The carbon atom of the >C–N–N< group from the hydrazone is believed to add to the carbanionic polymer chain, which after quenching with an active hydrogen-containing compound, such as water, an alcohol, an acid,

It has been demonstrated that styrene/butadiene interpolymers with a terminal functionalization can provide excellent combinations of physical properties (97).

3.14.6 *Liquid Polymer*

Liquid or low molecular weight polymers have been mixed with high molecular weight polymers in tire tread rubber compositions to impart good processability and to offer a better balance of wet and snow properties (98).

However, a significant drawback in using liquid polymers is their processing and handling in the manufacturing process. For example, liquid polymers may be made by polymerizing monomer units in a hydrocarbon solvent until a desired molecular weight is achieved. However, isolating the liquid polymer from the solvent by conventional methods, such as coagulation, is difficult due to the low molecular weight of the liquid polymer. To ease the ability of desolventizing the liquid polymer, the liquid polymer cement is often blended with a polymer cement of a higher molecular weight polymer, followed by desolventizing the blended cements. However, this procedure requires the liquid polymer and high molecular weight polymer to be polymerized separately and necessitates the additional step of blending. Thus, the processing efficiency is decreased.

A method has developed that uses (98):

1. Forming a living liquid polymer, wherein said living liquid polymer is anionically initiated and comprises a cation,
2. Adding a functional initiator precursor,
3. Adding monomer, wherein the functional initiator initiates the anionic polymerization of the monomer, and
4. Terminating the polymerization reaction initiated in the previous step.

Advantageously, this method may be performed in a single reactor. The liquid polymer may be produced by anionically polymerizing monomers capable of undergoing anionic polymerization upon reaction with an anionic initiator. Any monomer capable of anionic polymerization may be used. Exemplary monomers are collected in Table 3.8 and some are shown in Figure 3.55.

Any anionic initiator may be used to produce the liquid polymer. Exemplary initiators are summarized in Table 3.9.

The anionic polymerization of the liquid polymer is conducted in a hydrocarbon solvent. Suitable hydrocarbon solvents include

Table 3.8 Monomers for a liquid polymer (98).

Diene monomers	Aromatic monomers
1,3-Butadiene	Styrene
Isoprene	α-Methylstyrene
2-Ethyl-1,3-butadiene	*p*-Methylstyrene
2,3-Dimethyl-1,3-butadiene	Vinyl toluene
Piperylene(1,3-pentadiene)	Vinyl anthracene
2-Methyl-1,3-pentadiene	2-Vinylpyridene
3-Methyl-1,3-pentadiene	4-Vinylpyridine
4-Methyl-1,3-pentadiene	1-Vinylnaphthalene
2,4-Dimethyl-1,3-pentadiene	2-Vinylnaphthalene
1,3-Hexadiene	1-α-Methylvinylnaphthalene
1,2-Diphenyl-4-methyl-1-hexene	2-α-Methylvinylnaphthalene

Table 3.9 Anionic initiators (98).

Initiator	Initiator
n-Butyl lithium	*p*-Tolyllithium
4-Phenylbutyl lithium	4-Butylcyclohexyl lithium
Lithium dialkyl amines	Lithium dialkyl phosphines
Lithiumalkyl aryl phosphine	Lithium diaryl phosphines

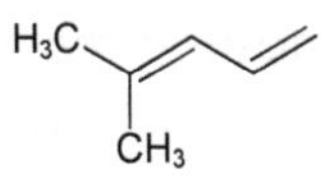

4-Methyl-1,3-pentadiene

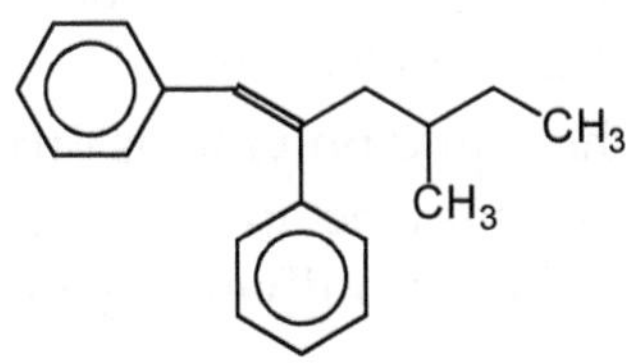

1,2-Diphenyl-4-methyl-1-hexene

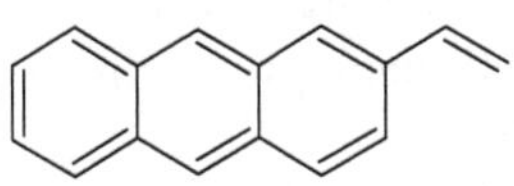

2-Vinyl anthracene

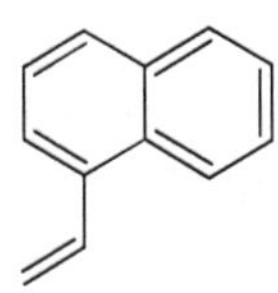

1-Vinylnaphthalene

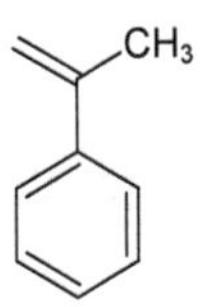

α-Methylstyrene

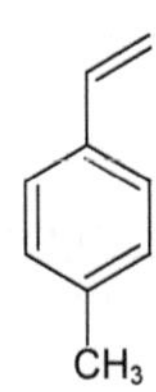

p-Methylstyrene

Figure 3.55 Monomers for a liquid polymer.

any suitable aliphatic hydrocarbons or alicyclic hydrocarbons. An alicyclic compound is an organic compound that is both aliphatic and cyclic.

Exemplary aliphatic hydrocarbons are pentane, isopentane, 2,2-dimethyl-butane, hexane, heptane, octane, nonane, decane. Exemplary alicyclic hydrocarbons are cyclopentane, methyl cyclopentane, cyclohexane, methyl cyclohexane, cycloheptane, cyclooctane, cyclononane, cyclodecane.

If the liquid polymer is produced using a diene monomer, a vinyl modifier may be added to increase the 1,2-addition reaction of the diene. Such modifying agents are shown in Table 3.10 and in Figure 3.56.

Table 3.10 Vinyl modifiers (98).

Compound	Compound
Ethylene glycol dimethyl ether	Diethylene glycol dimethyl ether
Triethylene glycol dimethyl ether	Tetraethylene glycol dimethyl ether
Tetrahydrofuran	1,4-Diazabicyclo[2.2.2]octane
Diethyl ether	Triethylamine
Tri-*n*-butylamine	Tri-*n*-butylphosphine
p-Dioxane	1,2-Dimethoxy ethane
Dimethyl ether	Methyl ethyl ether
Ethyl propyl ether	Di-*n*-propyl ether
Di-*n*-octyl ether	Anisole
Dibenzyl ether	Diphenyl ether
Dimethylethylamine	2,2-Di(2-tetrahydrofuryl)propane
Bis-oxalanyl propane	Tri-*n*-propyl amine
Trimethyl amine	Triethylamine
N,N-Dimethyl aniline	*N*-Ethylpiperidine
N-Methyl-*N*-ethyl aniline	*N*-Methylmorpholine
Hexamethylphosphoric acid triamide	*N,N,N',N'*-Tetramethylethylene diamine

The polymerization of the liquid polymer may be terminated when the desired number average molecular weight is reached, e.g., about 20 *k Dalton* to 100 *k Dalton*. The reaction is terminated by adding a functional initiator precursor. Suitable functional initiator precursors are shown in Table 3.11 and some compounds are shown in Figure 3.57.

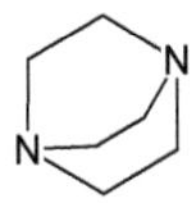

1,4-Diazabicyclo[2.2.2]octane

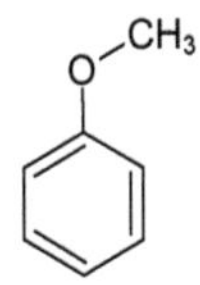

Anisole

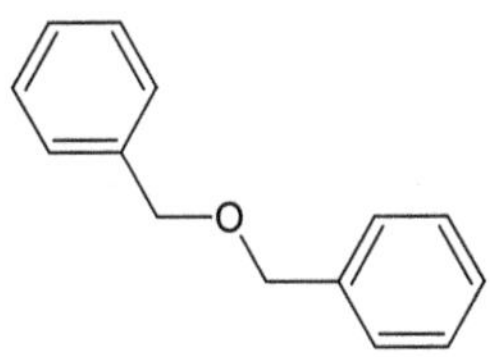

Dibenzyl ether

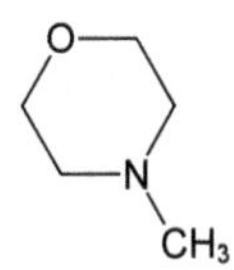

N-Methylmorpholine

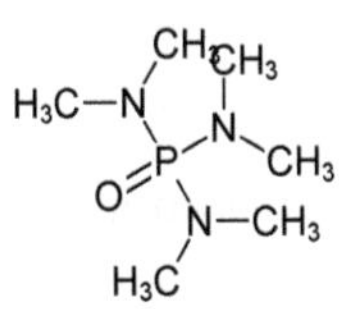

Hexamethylphosphoric acid triamide

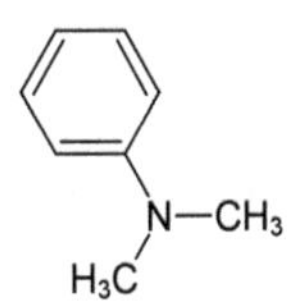

N,N-Dimethyl aniline

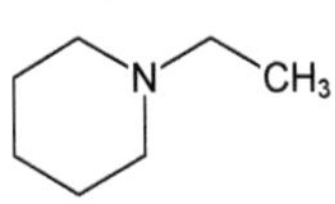

N-Ethylpiperidine

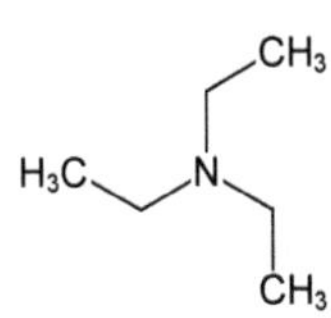

Triethylamine

Figure 3.56 Vinyl modifiers.

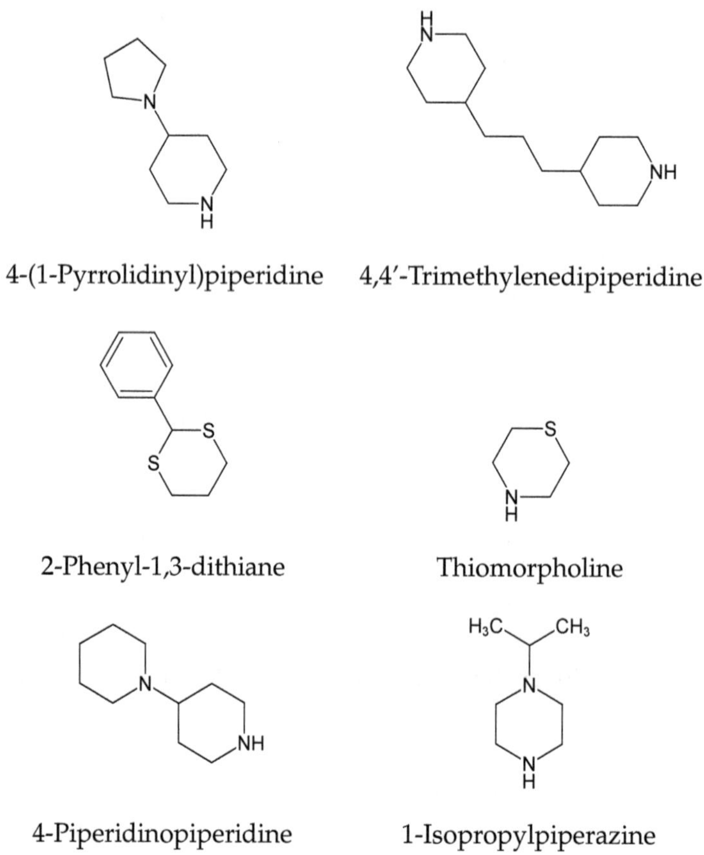

Figure 3.57 Functional initiator precursors (98).

Table 3.11 Functional initiator precursors (98).

Compound	Compound
Pyrrolidine	Piperidine
Hexamethylene imine	4-(1-Pyrrolidinyl)piperidine
4-Piperidinopiperidine	4,4′-Trimethylenedipiperidine
1-Isopropylpiperazine	1-(3-Methoxyphenyl)piperazine
1-(2-(Dimethylamino)ethyl) piperazine	1-[3-(dimethylamino) propyl]piperazine
Thiomorpholine	2-Phenyl-1,3-dithiane
2-[4-(Dimethylamino)] phenyl-1,3-dithiane	2-[4-(diethylamino)] phenyl-1,3-dithiane
2-[4-(4-Methylpiperazine) phenyl]-1,3-dithiane	Tributyltin hydride
Trioctyltin hydride	

If the functionalized polymer is polymerized in the presence of the liquid polymer according to the preparation process 3–13 described below, it is believed another advantage is that the microstructure of the liquid polymer and functionalized polymer will be more uniform.

Typically, these vulcanizable rubber compositions also include reinforcing fillers, such as carbon black and/or silica, and at least one vulcanizing agent. These compositions typically also include other compounding additives such as accelerators, oils, waxes, scorch inhibiting agents, and processing aids. These ingredients are known in the art, and may be added in appropriate amounts based on the desired physical and mechanical properties of the vulcanizable rubber composition. A liquid polymer can be synthesized as follows (98):

Preparation 3–13: First, 111.0 *g* of hexane, 100.0 *g* of 22.% 1,3-butadiene in hexane, 0.07 *ml* of bis-oxalanyl propane, and 0.23 *ml* of *n*-BuLi in hexane were added to a dried 28 ounce glass bottle that had been sealed with extracted septum liners and perforated crown caps under a positive nitrogen purge. The bottle was agitated and heated at 50°C for 1 *h* to form a living polymer cement. Then, 0.35 *ml* of piperidine solution (1.0 *M* in toluene) was added to terminate the living polymer cement, resulting in a piperidine lithium and a liquid polymer having a number average molecular weight of 54,747 $g\,mol^{-1}$, as determined by gel permeation chromatography using a PS standard, 200.0 *g* of 22.0% 1,3-butadiene in hexane was then added to the bottle. The bottle was agitated and heated at 50°C for 1 *h*. The polymerization was then terminated by adding 1.5 *ml* of isopropanol, resulting in a functionalized polymer having a M_n of 120,407 $g\,mol^{-1}$. The polymer cement was then treated with 3 *ml* of 2% di-*tert*-butyl-*p*-cresol in hexane (an antioxidant), coagulated in isopropanol, and drum dried. The product consisted of 67% of a functionalized polymer and 33% of a liquid polymer dispersed therein. The synthesis is shown in Figure 3.58.

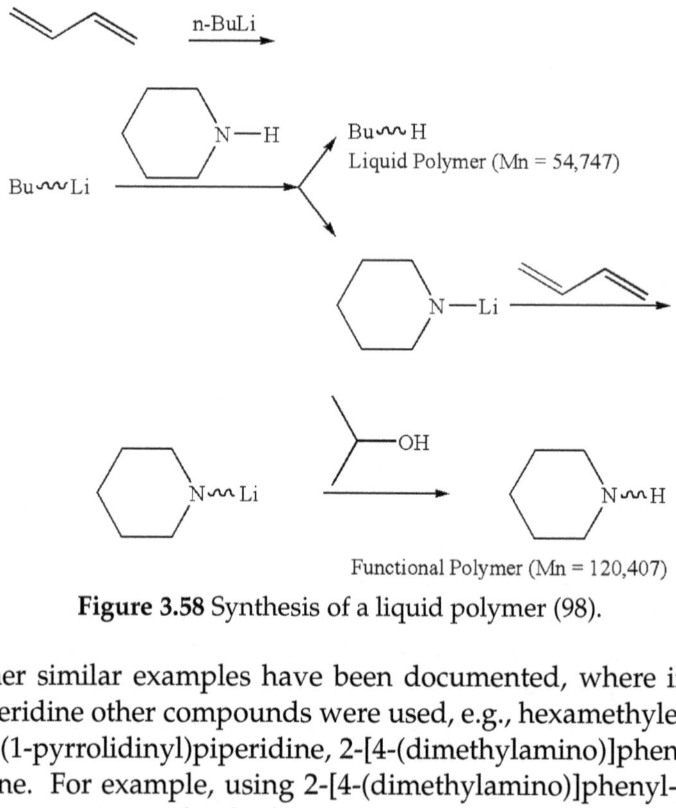

Figure 3.58 Synthesis of a liquid polymer (98).

Other similar examples have been documented, where instead of piperidine other compounds were used, e.g., hexamethylene imine, 4-(1-pyrrolidinyl)piperidine, 2-[4-(dimethylamino)]phenyl-1,3-dithiane. For example, using 2-[4-(dimethylamino)]phenyl-1,3-dithiane as terminator for the first-step polymerization and 1-methyl-2-pyrrolidone as terminator for the second-step polymerization, the reaction runs as shown in Figure 3.59.

The polymers synthesized were compounded according to the formulation shown in Table 3.12. In the formulation shown in Table 3.12, *N*-phenyl-*N'*-(1,3-dimethylbutyl)-*p*-phenylenediamine acts as an antioxidant while benzothiazyl-2-cyclohexylsulfenamide and *N,N'*-diphenyl guanidine act as accelerators (98). These compounds are shown in Figures 3.60 and 3.61.

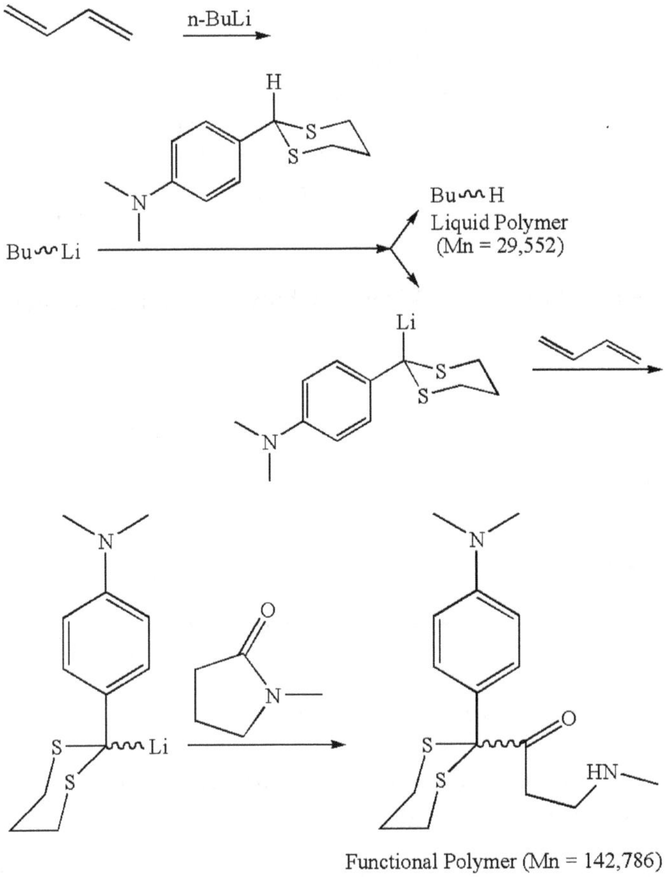

Figure 3.59 Synthesis of a liquid polymer (98).

Table 3.12 Formulation (98).

Compound	Amount/[*phr*]
Masterbatch Polymer	100
Carbon black (N343 type)	55
Wax	1
N-Phenyl-*N*′-(1,3-dimethylbutyl)-*p*-phenylenediamine	0.95
ZnO	2.5
Stearic acid	2
Aromatic processing oil	10
Final sulfur	1.3
Benzothiazyl-2-cyclohexylsulfenamide	1.7
N,*N*′-Diphenyl guanidine	0.2

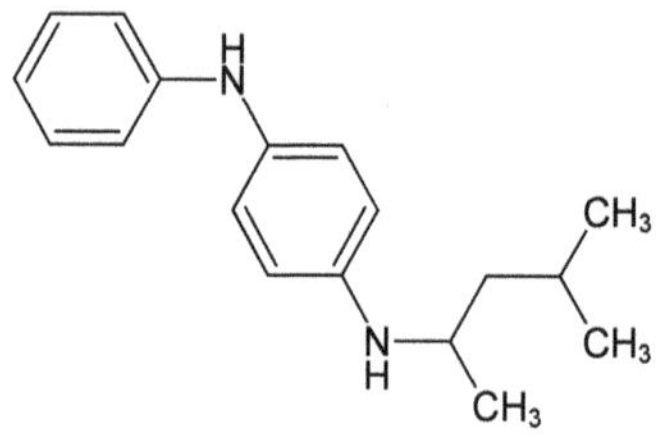

Figure 3.60 *N*-Phenyl-*N*′-(1,3-Dimethylbutyl)-*p*-phenylenediamine (98).

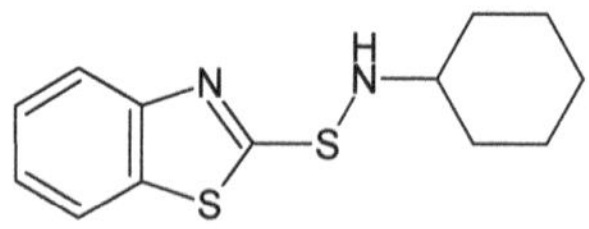

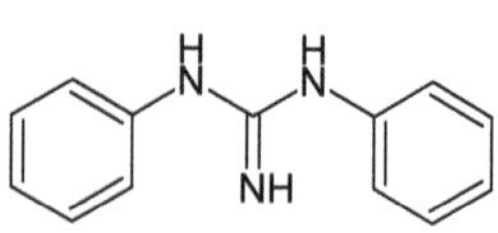

Benzothiazyl-2-cyclohexylsulfenamide *N*,*N*′-Diphenyl guanidine

Figure 3.61 Accelerators (98).

3.14.7 Polycyano-Functionalized Polymers

The preparation of a vulcanizable composition has been reported that consists of the steps of (99):

1. Polymerizing a conjugated diene monomer and optionally vinyl-substituted aromatic compounds to thereby form a reactive polymer having a reactive chain end, and
2. Reacting the polymer with a polycyano compound, a filler, and a curative.

A series of preparation examples for the modification of *cis*-1,4-poly(butadiene) and poly(styrene-*co*-butadiene) with cyano compounds have been detailed. The cyano compounds used therein are shown in Table 3.13 and in Figure 3.62.

Also, poly(butadiene) polymers with a protected oxime compound containing a cyano group have been described (100). Here, 2-cyanobenzaldehyde *o*-methyloxime, c.f. Figure 3.63, and 4-cyanobenzaldehyde *o*-methyloxime were used, among other related compounds.

Table 3.13 Cyano compounds for modification (99).

Compound	Compound
1,2-Dicyanobenzene	1,2-Dicyanoethane
1,3-Dicyanopropane	Adiponitrile
Benzylidenemalononitrile	Fumaronitrile
Benzonitrile	Acetonitrile
1,2-Dicyanobenzene	Adiponitrile

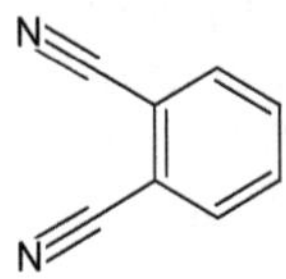

1,2-Dicyanobenzene

1,2-Dicyanoethane

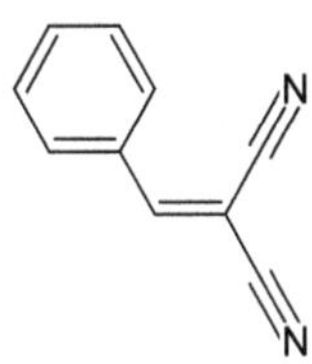

Benzylidenemalononitrile

Fumaronitrile

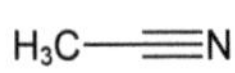

Acetonitrile

Adiponitrile

Figure 3.62 Cyano compounds (99).

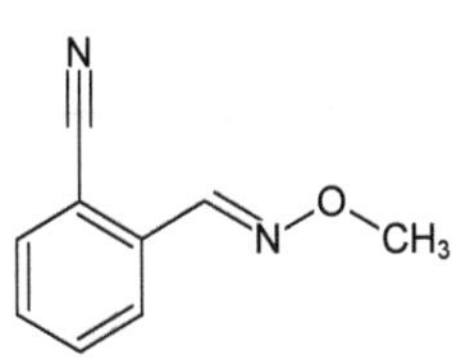

Figure 3.63 2-Cyanobenzaldehyde *o*-methyloxime.

3.14.8 *Functionalized Polymer with a Protected Amino Group*

A functionalized polymer that contains a heterocyclic protected amino group has been described (101). The protected amino group is selected from the group consisting of bis(trihydrocarbylsilyl)amino, bis(dihydrocarbylhydrosilyl)amino, 1-aza-disila-1-cyclohydrocarbyl, (hydrocarbyl)(trihydrocarbylsilyl)amino, (hydrocarbyl)(dihydrocarbylhydrosilyl)amino, 1-aza-2-sila-1-cyclohydrocarbyl, dihydrocarbylamino, and 1-aza-1-cyclohydrocarbyl groups.

Specific examples are 2-[bis(trimethylsilyl)amino]pyrimidine, [bis(trimethylsilyl)amino]pyrazine, [bis(trimethylsilyl)amino]-1,4-triazine, and [bis(trimethylsilyl)amino]-1,3,5-triazine. Some specific compounds used here are shown in Figure 3.64.

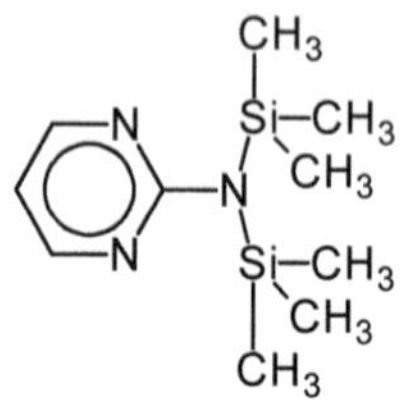

2-[Bis(trimethylsilyl)amino]pyrimidine

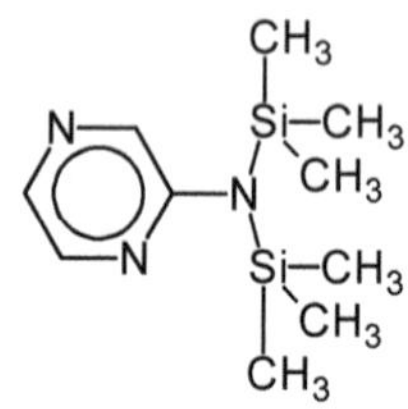

[Bis(trimethylsilyl)-1,4-amino]pyrazine

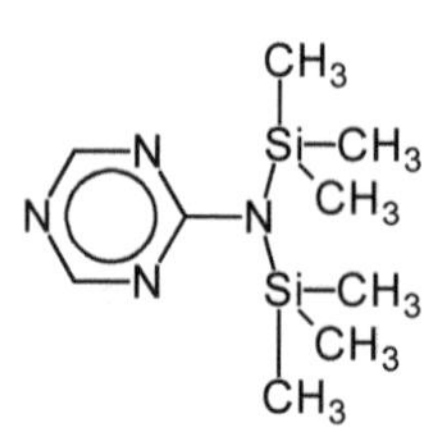

[Bis(trimethylsilyl)-1,3,5-amino]pyrazine

Figure 3.64 Protected amino groups (101).

The synthesis of [Bis(trimethylsilyl)amino] can be done as follows (101):

Preparation 3–14: About 4.84 *g* of aminopyrazine, 11.33 *g* of triethylamine, and 10 *ml* of toluene were mixed in a round-bottom reaction flask cooled with an ice bath. To this mixture was added, in a dropwise fashion, a solution of 24.84 *g* of trimethylsilyl trifluoromethanesulfonate in 60 *ml* of toluene. The resulting mixture was stirred at room temperature for 3 *d* to give a biphasic mixture. The top layer was transferred to another flask, and the bottom layer was extracted with 40 *ml* of toluene. The combined toluene solution was evaporated under vacuum. The residue was extracted with 100 *ml* of hexane, and the hexane layer was evaporated under vacuum, yielding [bis(trimethylsilyl)amino]pyrazine as a yellow oil with 11.74 *g*, i.e., a yield of 96%.

The functional group in the functionalized polymer may reduce the number of free polymer chain ends via interaction with filler particles. Also, the functional group may reduce the agglomeration of the filler.

3.14.9 Comb Block Copolymers

Halobutyl rubbers, which are halogenated isobutylene/isoprene copolymers, are the polymers of choice for best air retention in tires for passenger, truck, bus, and aircraft vehicles (102). Bromobutyl rubber, chlorobutyl rubber, and halogenated star-branched butyl rubbers can be formulated for specific tire applications, such as tubes or innerliners.

In addition, these elastomers are preferred barrier layers for pharmaceutical stoppers and are useful as engine mounts for damping purposes. Examples of these elastomers are brominated isobutylene-isoprene rubber, chlorinated isobutylene-isoprene rubber, or a brominated isobutylene-*co*-*p*-methyl-styrene copolymer.

However, the inertness of saturated hydrocarbon polymers, their low reactivity and incompatibility with most other materials, and the difficulties in adhering them to, or using them in conjunction with most other materials, has restricted their use in many areas.

For example, halobutyl rubbers are incompatible with general purpose rubbers, such as SBR, natural rubber, and *cis*-butadiene rubber. They are also incompatible with acrylics, graphene, graphite, carbon blacks, and silica due to their inherent hydrophobicity.

These incompatibilities prevent their use in tire tread compounds for better damping and traction, in acrylic latex coating for better impermeability and less moisture sensitivity, and in nanocomposites for lower permeability and better mechanical properties.

Methods have been developed for improving the compatibility of halobutyl rubbers with general purpose rubbers, acrylics, and fillers such as graphite, graphene, carbon black, and silica (102).

Comb block copolymers were formulated that have a halobutyl rubber backbone and functional polymer comb arms. The functional polymer comb arms improve the compatibility of these comb block copolymers with other materials. The halogenated copolymer, which forms the backbone for the comb block copolymers, is a copolymer of an isoolefin having from 4 to 7 carbon atoms and a *p*-alkylstyrene.

The cleavage of the carbon halogen bond in the benzylic-halide moiety in these molecules can be readily used to initiate the ATRP of acrylates, styrenes, acrylamides, and/or acrylonitriles to produce the comb block copolymers (102)-

3.15 Polymer Composition for Grease

Greases can be prepared from base oil, an overbased calcium sulfonate thickener, as well as conventional thickener, such as lithium 12-hydroxy stearate and optionally other performance additives, e.g., antioxidants or antiwear agents (103,104).

A grease containing an overbased calcium sulfonate thickener is known to have acceptable corrosion inhibiting properties. However, this type of grease has poor water-resistance properties, such as poor water wash-off or water repellency, as can be seen in an ASTM D4049 water spray-off test (105). Grease with poor water wash-off or water repellency decreases the longevity of grease and increases wear on the surface being lubricated.

Polymers have also been added to a grease composition other than overbased calcium sulfonate grease to improve the performance characteristics of the grease. For example, polymers have been employed to decrease water wash-off, to decrease oil separation, to increase water repellency, to increase dropping points or cone penetration and as thickeners. The polymers are poly(methacrylate)s or polyolefins. Typically, these polymers are incorporated into the base oil and act as a viscosity modifier. However, the polymers have limited interaction with the thickener, resulting in the grease being more susceptible to the effects of water such as water wash-off or decreased water repellency.

Therefore, it would be desirable to have a grease that is capable of imparting improved thickening. A lubricating grease composition has been developed that contains (103):

1. The reaction product of a calcium-containing overbased organic acid,
2. At least one acid-producing compound, and
3. An oil of lubricating viscosity.

A functionalized polymer can be used that contains an unsaturated dicarboxylic acid anhydride or derivatives thereof. The functionalized polymer containing an unsaturated dicarboxylic acid anhydride or derivatives thereof may be used alone or in combination. The functionalized polymer includes a grafted functionalized polyolefin, an esterified polymer, and a polymer derived from carboxylic acid ester monomers.

The preparation of a functionalized polymer can be done as follows (106):

Preparation 3–15: To a 5 *l*, four-necked flask equipped with a stirrer, nitrogen inlet, subsurface tube, thermowell, and condenser, is charged 2121 *g* of stock mineral oil (#151). The oil is stirred and heated to 160°C under a nitrogen flow of about 8 $l\,h^{-1}$. To the flask is added 374.3 *g* LZ® 7060 ethylene-propylene/dicyclopentadiene polymer, number average molecular weight about 115,000 *Dalton*, in the form of 1 *cm* cubes, over the course of about 1 *h*. The mixture is thereafter stirred for an additional 2 *h* and allowed to cool overnight.

Then, the mixture is heated under nitrogen to 160°C with stirring for 3 *h*. Next, 3.8 *g* maleic anhydride is added and the mixture is stirred for an additional 15 *min*. To the mixture is added 3.8 *g* di-*tert*-butyl peroxide, dropwise, over 1 *h* at 160°C. The temperature is maintained for an additional 2.5 *h*. Thereafter the nitrogen flow is increased to 42 $l\,h^{-1}$ for 1.5 *h*. Then, the flask is cooled and sealed overnight. The mixture is heated under nitrogen for an additional 3 *h* at 160°C. Upon cooling the product (without further isolation) is maleic anhydride functionalized olefin copolymer in oil.

Other suitable examples of the unsaturated dicarboxylic acid anhydride include maleic anhydride, methyl maleic anhydride, ethyl maleic anhydride, dimethyl maleic anhydride, c.f. Figure 3.65, or mixtures thereof (103).

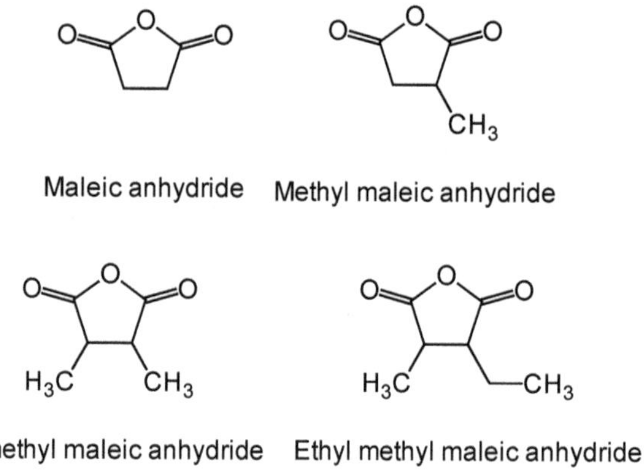

Figure 3.65 Maleic acid derivatives.

3.16 Hydrogels

3.16.1 Tough Hydrogels Crosslinked by Multifunctional Polymer Colloids

Polymer hydrogels are three-dimensional polymer networks with a very high water content. Polymer hydrogels have great potential applications in soft actuators, artificial muscles, tissue engineering, and others (107).

A comprehensive account of most of the recent progress on tough hydrogels crosslinked by polymer colloids, which explores the toughening mechanisms, has been presented (107). The fields of application are summarized in Table 3.14.

To improve the strength and toughness of hydrogels, numerous strategies have been developed to integrate efficient energy dissipation mechanisms into the hydrophilic networks. Among them, the use of macro-crosslinkers to replace conventional chemical ones has become promising to develop tough hydrogels. Polymer colloids, including nano-/microparticles, nanogels and microgels, hydrophobic associates, and block copolymer assemblies, have been employed as multifunctional macro-crosslinkers that link the polymer chains through covalent bonds or noncovalent interactions.

The dislocation, deformation, dissociation, and rupture of poly-

Table 3.14 Applications of polymer hydrogels (107).

Application	References
Tissue engineering	(108, 109)
Adhesives	(110, 111)
Drug delivery	(112, 113)
Microlenses	(114)
Sensors	(115)
Artificial muscles	(116, 117)
Soft robotics	(118)
Energy storage and conversion	(119, 120)
Electrochemical devices	(121)

mer colloids upon loadings are the major mechanisms to dissipate energy.

A toughening concept has been used for neutral, biocompatible, and biofunctional polymers (122, 123) for applications in cartilage repair (124, 125).

Silica nanoparticles have been widely used to reinforce polymer hydrogels (126–130). The abundant hydroxyl groups on silica surface and the extremely high specific surface are beneficial for the chain adsorption. The strong adsorption of hydrophilic poly(acrylic acid) (131) or poly(*N*,*N*-dimethylacrylamide) (132) chains onto silica nanoparticles at high concentrations of around 20% results in a strong adhesive that glues tissues together.

Nanohybrid hydrogels made from hyaluronic acid and silica nanoparticles with controllable modulus (133) are used to regulate the preferential differentiation of stem cells (134).

On the other hand, surface functionalized silica nanoparticles have been employed to copolymerize with hydrophilic monomers. The polymer chains are bonded and adsorbed on the nanoparticle surface to form a core-shell structure (129, 135).

Moreover, grafting a functional poly(butyl acrylate) layer on silica nanoparticle surface results in inorganic-organic core-shell hydrophobic latexes that can adsorb hydrophobic lauryl methacrylate monomers for *in-situ* copolymerization with acrylamide, generating hydrogels with outstanding stretchability, fatigue resistance, and self-recoverability (136).

Upon loading, the particles are able to dislocate to redistribute the stress. At high strain, the particle-particle distance is enhanced along the elongation, whereas that normal to stretching is decreased. The significant changes in particle-particle correlation also dissipate energy. The rigid particles, however, are not able to deform. Most silica nanoparticle crosslinked hydrogels show an elastic behavior. To synthesize viscoelastic hydrogels, dynamic interactions in crosslinks or between polymer chains are needed.

In comparison with conventional small-molecule crosslinking agents, e.g., *N,N'*-methylenebis(acrylamide), polymer colloids are much larger and can be conveniently functionalized on the surface or in bulk to participate in polymerization, and covalently bond and crosslink the polymer chains (137–140).

Hydrogels crosslinked by polymer nanoparticles or microparticles show very high extensibility and resilience (141–143).

Multifunctional nanoparticles-hydrogel nanocomposites made with chitosan hydrogel beads and solid lipid-polymer hybrid nanoparticles were prepared through the conjugation between solid lipid-polymer hybrid nanoparticles and chitosan beads (144). The solid lipid-polymer hybrid nanoparticles were first fabricated via coating the bovine serum albumin-emulsified solid lipid nanoparticles with oxidized dextran. The aldehyde groups of the oxidized dextran on the surface of the solid lipid-polymer hybrid nanoparticle enabled an *in-situ* conjugation with the chitosan beads through the Schiff base linkage.

The obtained nano-on-beads composite exhibited a spherical shape with a homogeneous size distribution. The successful conjugation of solid lipid-polymer hybrid nanoparticle on the chitosan beads was confirmed by FTIR and SEM. The effects of the beads dosage and the incubation duration on the conjugation efficiency of the solid lipid-polymer hybrid nanoparticles onto the beads were comprehensively optimized.

The optimal formulations were found to be a 200 bead dosage, with 30–90 *min* incubation duration groups. These optimal formulations were then used to encapsulate thymol, an antibacterial agent, which was studied as a model compound. After encapsulation, the thymol exhibited sustained release profiles in the phosphate buffer saline. The so-prepared nanoparticles-hydrogel nanocomposites show promising features as a controlled-release antibacterial approach for improving food safety (144).

At the moment, there are no therapies that can ameliorate the neurological impairments of spinal cord injury, and the long-term spinal cord injury survival rate is far from satisfactory (145). Electrical stimulation and electrically conductive conduits are emerging as promising strategies with positive results showing increased neurite and axon growth, both *in vitro* and *in vivo*.

Carbon nanotubes were functionalized with hydrophilic poly(ethylene glycol) chains and further crosslinkablility was introduced with double bonds (145). The final carbon-nanotube-poly(ethylene glycol)-acrylate material was embedded within oligo(poly(ethylene glycol) fumarate) using varying concentrations to form conductive hydrogels with modulable conductivities.

In-vitro neural cell adhesion and proliferation, as well as neurite development after nerve growth factor induction, showed a significant enhancement in the conductive hydrogels. Various types of nerve conduits were then successfully fabricated using an injection molding technique. With many desired properties, these conductive nerve conduits have promising potential for promoting axon guidance and enhancing nerve recovery in spinal cord injury patients (145).

In summary, multifunctional polymer colloids including polymer particles, nano-/microgels, hydrophobic associates, or copolymer micelles, have been widely used as macro-crosslinking agents to prepare tough hydrogels (107). These unique soft macro-crosslinking agents at the length scale from a few nanometers to submicrometers can form either covalent bonds or noncovalent interactions with polymer chains.

Only a little energy dissipation occurs during deformations as most glassy polymer particles are not able to deform upon loading. By establishing a hydrophobic association between polymer chains with polymer particles, the composite hydrogels gain outstanding energy dissipation properties due to the reversible detachment of polymer chains from the particle surface.

In contrast to the rigid glassy polymer particles, soft polymer colloids like nano-/microgels as macro-crosslinkers of bulk hydrogels are able to deform upon loading, and thus improve the extensibility and toughness of hydrogels. The permeability of microgels even enables the formation of double network structures in the local domains (107).

3.16.2 *Aptamer-Functionalized DNA Hydrogel*

An aptamer-integrated deoxyribonucleic acid (DNA) hydrogel can be used as a protein delivery system with an adjustable release rate and time by using complementary sequences as the biomolecular trigger (146).

This can be done with a simple functionalization method. The aptamer-functionalized DNA hydrogel was prepared via a one-pot self-assembly process from two kinds of DNA building blocks (X-shaped and L-shaped DNA units) and a single-stranded aptamer.

The gelling process was achieved under physiological conditions within 1 *min*. In the absence of the triggering complementary sequences, the aptamer grafted in the hydrogel exhibited a stable state for protein-specific capture. While hybridizing with the triggering complementary sequences, the aptamer is turned into a double-stranded structure, resulting in the fast dissociation of protein with a wise-stage controlled release program.

Furthermore, the DNA hydrogel with an excellent cytocompatibility could be successfully applied to human serum, forming a complex matrix. The whole process of protein capture and release was biocompatible and could not refer to any adverse factor of the protein or cells. Thus, the aptamer-functionalized DNA hydrogel will be a good candidate for controlled protein delivery (146).

3.16.3 *Macrocyclic Hydrogel System*

Macrocyclic hydrogel systems were synthesized using sol-gel polymerization for high selectivity adsorption materials. The hydrogel was fabricated from maleic acid and ethylene diamine. The synthesis is shown in Figure 3.66. It could be shown that 1,4-diazocane-5,8-dione were the principal active sites of this adsorbent (147).

The macrocyclic sites provided defined zones which limited capture and chelation of transition metal ions such that from single ion solutions only ion substrates of between 1.23 Å and 1.29 Å ionic size were mainly adsorbed. In the case of metal-metal competitive solutions, only 1.29 Å sized substrates would be adsorbed.

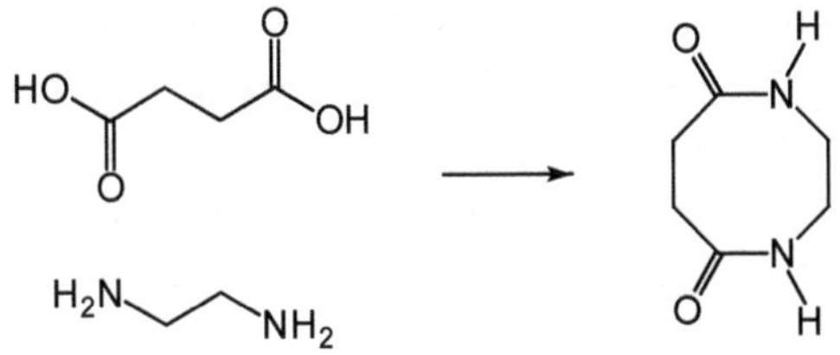

Figure 3.66 Synthesis of the bismalimide pre-product 1,4-diazocane-5,8-dione (147).

This adsorption specificity required that metals dissolved in water of these sizes must be of octahedral configurations for more optimum host-guest chelation with the gel active sites. Since Fe(aq) effectively met these criteria, the gel adsorbent displayed outstanding specificity for Fe(aq) adsorption from different levels of competitive environments. The adsorption kinetics of several metals was analyzed using a pseudo-second-order equation:

$$\frac{1}{q} = \frac{1}{q_e}t + \frac{1}{k_2 q_e^2} \tag{3.1}$$

Here, t is the absorption time, q_e is the equilibrium adsorption capacity and $k_2$is the rate constant for the second-order kinetic equation. According to this model, adsorption occurs via chemisorption if in a plot of t/q vs. t a highly linear fit would be obtainable. The results of these calculations are collected in Table 3.15.

Table 3.15 Kinetic model analysis of maleic acid-ethylene diamine hydrogel adsorption behavior (147).

Metal ion	Experimental	Calculated pseudo-second-order equation kinetic analysis		
	q_e	q_e	k_2	R^2
Cr	21.22	29.85	0.0022	0.9080
Fe	33.19	33.67	0.0782	0.9856
Cu	15.41	16.50	0.0139	0.9945
Ni	3.57	3.92	0.1435	0.9915
Co	4.62	4.55	0.2916	0.9816

The adsorbent showed that macrocyclic chelate sites can be infused directly into the hydrogel network without any grafting, and facilitates a shape-based, size-limited adsorption. So, this hydrogel is a promising candidate for applications in heavy metal pollution remediation, drug development and fuel cell catalysts, where high specificity adsorption properties are desirable (147).

3.16.4 *Redox-Responsive Hydrogels*

Nanophase-separated amphiphilic polymer networks are ideally suited as responsive membranes due to their stable cocontinuous structure (148). Their functionalization with redox-responsive 2,2′:6′,2″-terpyridine-metal complexes, c.f. Figure 3.67, and light-responsive spiropyran derivatives results in a material with tunable optical, redox and permeability properties.

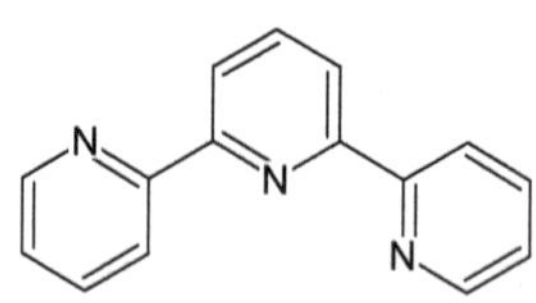

Figure 3.67 2,2′:6′,2″-Terpyridine.

The versatility of the system in complexing various metal ions, such as cobalt or iron at different concentrations, results in a perfect monitoring of the degree of crosslinking of the hydrophilic poly(2-hydroxyethyl acrylate) channels.

The reversibility of the complexation, the redox state of the metal and the isomerization to the merocyanine form, c.f. Figure 3.68, upon UV illumination was evidenced by cyclic voltammetry, UV-Vis and permeability measurements under sequential conditions. Thus, such membrane types provide light and redox addressable functionalities due to their adjustable and mechanically stable hydrogel network (148).

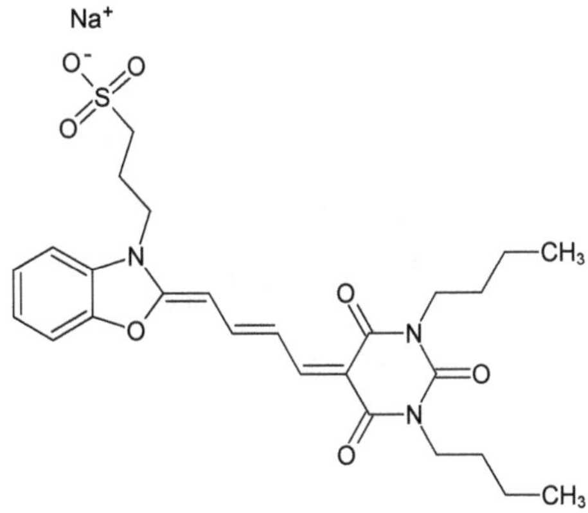

Figure 3.68 Merocyanine.

3.17 Coordinating Polymers

Coordinating polymers, i.e., polymers functionalized with organic ligands that enable metal ion coordination, have found applications as scavenger reagents and as catalyst supports in organic synthesis (149–151). Other uses include their employment as antimicrobial materials for wastewater treatment and for the recovery of trace metal ions (152–154)

A one-pot method for the functionalization of poly(vinylbenzyl chloride) with a tris(benzyltriazolylmethyl)amine, c.f. Figure 3.69, ligand via the copper-catalyzed azide-alkyne cycloaddition reaction has been reported (155). Here, the ligand is constructed simultaneously with its attachment to the polymer backbone. A hydrophobic tris(benzyltriazolylmethyl)amine polymer, as well as copolymers containing tris(benzyltriazolylmethyl)amine ligands, were also prepared to demonstrate the possibility of tailoring the properties of the polymers.

The polymers were characterized with FTIR and NMR spectroscopy, elemental analysis, contact angle measurement, DSC and TG. The metal coordinating properties of films prepared from the polymers were also demonstrated using X-ray photoelectron spectroscopy and their structures were inspected by SEM (155).

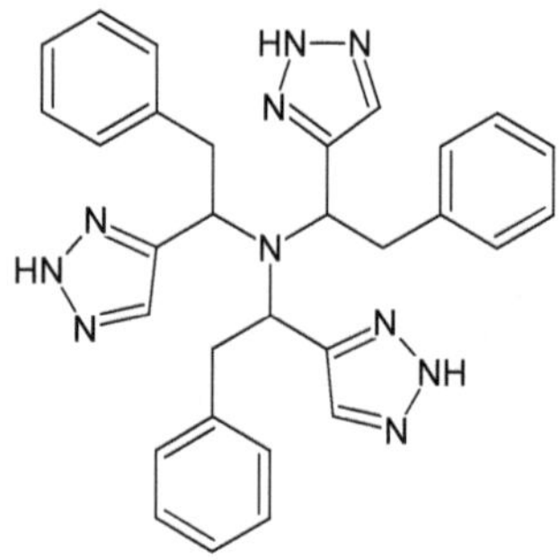

Figure 3.69 Tris(benzyltriazolylmethyl)amine.

3.18 Dye Removal

3.18.1 *Macrocyclic Functionalization*

Crosslinked PS beads were initially functionalized with phenacyl ester linkers, and a series of four macrocycles with different ring sizes were synthesized from these linkers (156). The macrocycles were built from amide-linked monomers coupled by conventional peptide synthesis methods. Annulation was achieved by copper-(I)-catalyzed intramolecular azide-alkyne cycloaddition to give triazole-linked macrocycles. The macrocycles were cleaved from the polymer beads using hydrazinolysis or saponification. The structures of macrocycles were confirmed by high-resolution NMR and liquid chromatography-mass spectroscopy analysis.

The ability of the polymers to selectively bind compounds from a mixture of aromatic derivatives in ethanol was tested. The prepared PS-supported macrocycles were found to selectively bind bromophenol blue and bromocresol green, c.f. Figure 3.70, noncovalently with an association constant of 160–490 M^{-1} (156).

3.18.2 *Cationic Dye Removal*

3.18.2.1 *Methyl Violet and Malachite Green*

A magnetic nanocomposite was synthesized via radical polymerization of methyl acrylate onto modified magnetic nanoparticles

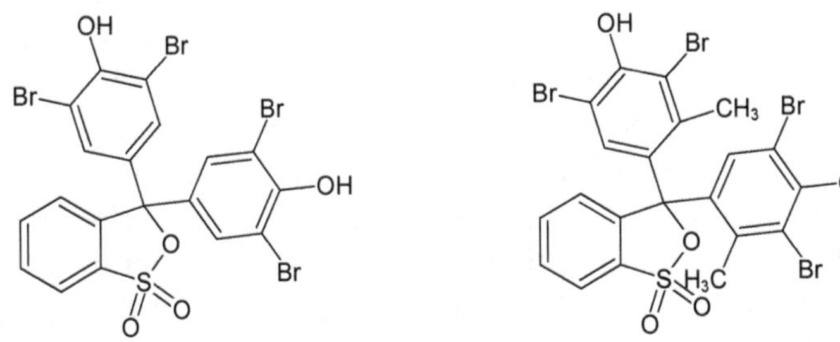

Bromophenol blue   Bromocresol green

Figure 3.70 Dyes.

followed by the functionalization of the methyl ester groups with ethylene diamine and sodium chloroacetate (157).

This product played a key role as an adsorbent for the removal of typical cationic dyes, methyl violet and malachite green, c.f. Figure 3.71.

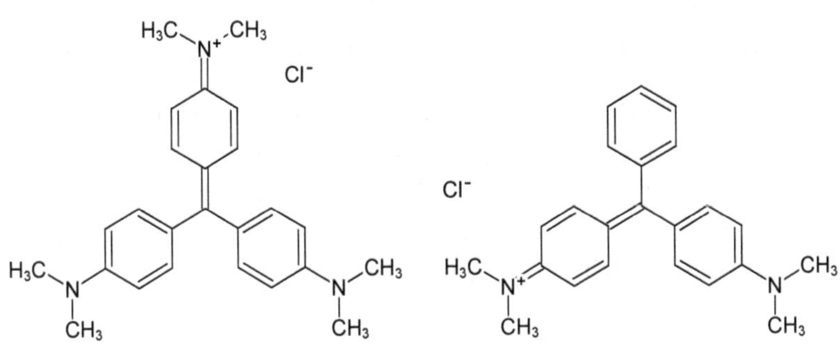

Methyl violet   Malachite green

Figure 3.71 Cationic dyes.

The dependence of the properties on pH, contact time and initial dye concentration was investigated (157). The adsorbent displays excellent adsorption capacities for cationic dyes. A study of the kinetic and isotherm of adsorption indicated that the dye adsorption process can be well described by pseudo-second-order kinetic models and Langmuir isotherm models, respectively.

In order to find an accurate correlation between the adsorption enthalpy ΔH and the experimental maximum adsorption capacity q_m, the calculation of the adsorption enthalpy of a series of cationic dyes onto the adsorbent was investigated using density functional theory (157). Density functional theory is a computational quantum mechanical modeling method used in physics, chemistry and materials science to investigate the electronic structure of many-body systems, in particular atoms, molecules, and the condensed phases (158).

An excellent agreement between the experimental q_m and the calculated ΔH values was found, which is able to predict the maximum adsorption capacities on the prepared adsorbent for other cationic dyes (157).

3.18.2.2 Reactive Blue 21

Chitosan/poly(amidoamine) microparticles were prepared as magnetic adsorbents for the removal of the Reactive Blue 21 dye from an aqueous solution (159). Reactive Blue 21 is shown in Figure 3.72.

The characterization of these particles was carried out using SEM, FTIR, X-ray diffraction diffractometry and vibrating sample magnetometry. The results indicated that the magnetic chitosan microparticles were functionalized with poly(amidoamine) dendrimers and maintained their intrinsic magnetic properties.

Also, the effects of initial pH, adsorbent dose, initial concentration, contact time and temperature on adsorption were investigated (159). Kinetic studies showed that the dye adsorption process followed a pseudo-second-order kinetic model. however, the adsorption rate was also influenced by intraparticle diffusion. Equilibrium adsorption isotherm data indicated a good fit to the Langmuir isotherm. The maximum adsorption capacities obtained from the Langmuir model are shown in Table 3.16.

The thermodynamic parameters revealed the feasibility, spontaneity and endothermic nature of the adsorption. Recycling experiments confirmed the reusability of the adsorbent (159).

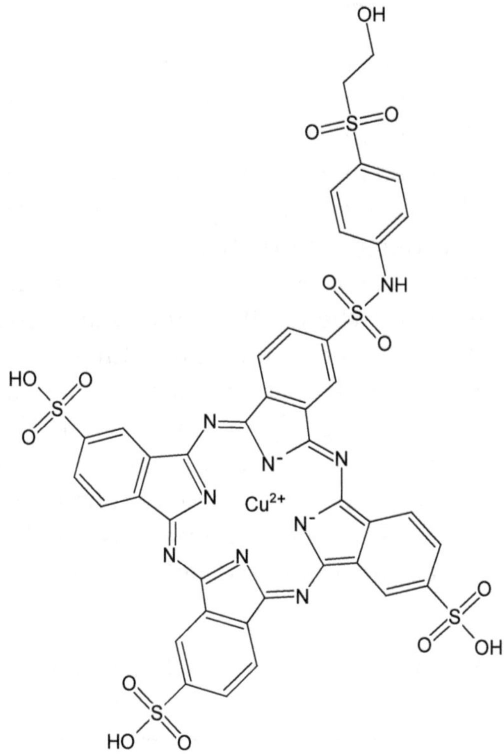

Figure 3.72 Reactive Blue 21.

Table 3.16 Maximum adsorption capacities (159).

Temperature K	°C	Maximum adsorption capacity/[$mg\,g^{-1}$]
303	30	555.56
313	40	588.24
323	50	625.00
333	60	666.67

3.19 Separation Processes

3.19.1 *Cellulose Adsorbing Agent*

A chemical product has been described, which can be used as a cellulose adsorbing agent or a crosslinking agent in the manufacture of cellulose products (160).

It effects an improved folding endurance, improved tensile index, tensile stiffness, tensile energy absorption, strain at break, modulus of elasticity or wet strength properties.

Here, a polymer is made of primary amine-functionalized polymer and hemicellulose, wherein the primary amine-functionalized polymer is covalently bound to the hemicellulose.

A hemicellulose can be any of several heteropolymers present in almost all plant cell walls, e.g., xylan, arabinoxylan, glucuronoxylan, glucuronoarabinoxylan, glucomannan, galactomannan, galactoglucomannan, and xyloglucan. Hemicellulose contains many different sugar monomers. Some of these compounds are shown in Figure 3.73. Primary amine-functionalized polymers are shown in Table 3.17.

Table 3.17 Primary amine-functionalized polymers (160).

Polymer type	Polymer type
Poly(allylamine)	Poly(vinylamine)
Poly(vinylamine-*co*-vinylformamide)	Chitosan
Poly(ethylene imine)	

An exemplary polymer may be made of chitosan and xyloglucan. The chitosan is covalently bound to xyloglucan. The chitosan and xyloglucan may be bound to each other by reductive amination between free amino groups of the chitosan and reducing end of xyloglucan.

It has been found that when xyloglucan is aminated with chitosan it forms a reaction product which, when used as a crosslinking agent in papermaking, confers superior folding endurance strength to the paper product obtained (160).

A polymer made of primary amine-functionalized polymer and hemicellulose, e.g., chitosan/xyloglucan material, may also be pro-

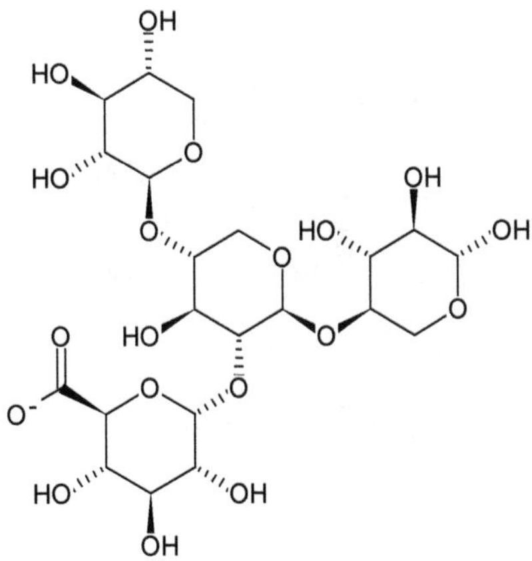

Glucuronoxylan

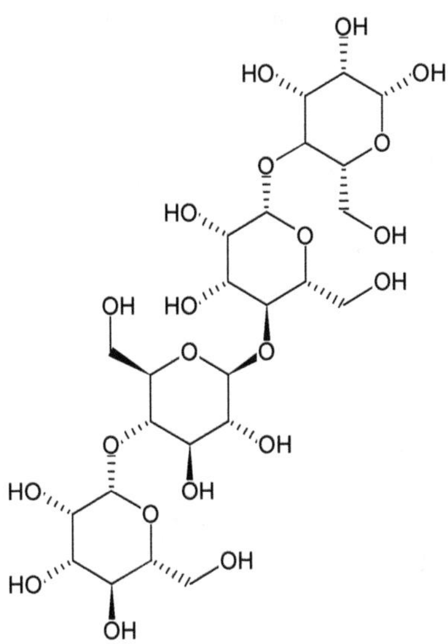

Glucomannan

Figure 3.73 Hemicellulose compounds.

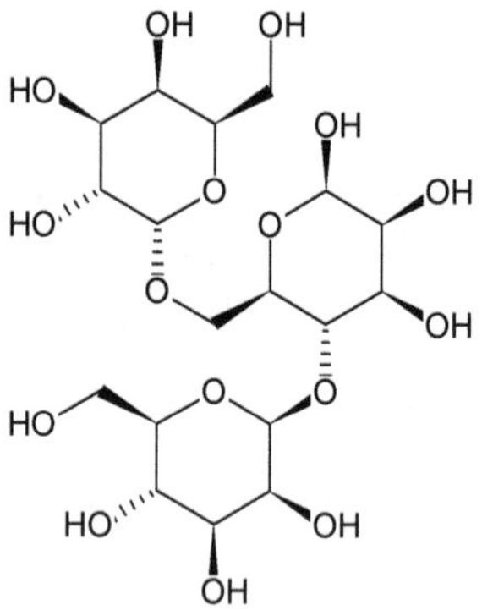

Galactomannan

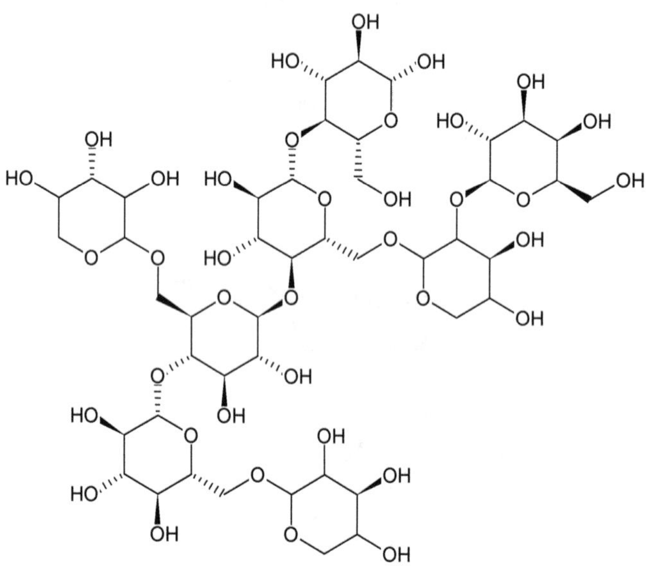

Xyloglucan

Figure 3.73 (cont.) Hemicellulose compounds

duced by the reductive amination involving the conversion of a carbonyl group to an amine via an intermediate imine. The reductive amination of carbohydrates with a reducing end can be performed in one pot, with the imine formation and reduction occurring concurrently (160).

3.19.2 *Amine-Functionalized Poly(styrene)*

Functionalized polymer particles are useful for chromatography and other separation processes as the solid phase for solid-phase organic synthesis, particularly synthesis of oligopeptides, oligonucleotides and small organic molecules, e.g., in combinatorial chemistry, and as supports for catalysts and reagents, e.g., for diagnostic assays (161).

Generally the preparation of amine-functionalized particles has involved the copolymerization of two or more monomers, one of which has a functional group which is transformable to an amine group or to which an amine group may be coupled after polymerization is complete. As a result the distribution of the amine groups throughout the amine-functionalized particle is generally non-uniform and less than optimal.

It has been found that amine-functionalized polymer particles can be prepared directly by suspension polymerization of aminostyrene, preferably 4-aminostyrene, together with at least one further vinylic monomer, especially a styrenic monomer, and, optionally, a crosslinking agent (161).

The suspension polymerization process is a seeded suspension polymerization in which the seed is swollen before the polymerization reaction is initiated and in which continuous or batchwise monomer addition continues during the suspension polymerization phase. Seed swelling has been described in the literature (162, 163).

The capacity of the seeds to swell could be increased to a volume increase of 125 times or even more if an organic compound with relatively low molecular weight and low water solubility is diffused into the seeds before the bulk of the monomer is used to swell the seeds. The effect is based on entropy rather than on the chemical nature of the organic compound in particular. Conveniently, the polymerization initiator may be used for this purpose. Organic solvents, e.g., acetone or a portion of the monomer, may be used to enhance the diffusion of the organic compound into the seeds. This *Ugelstad polymerization process* (162) may be used to produce monodisperse particles, if necessary carrying out several swelling and polymerization stages to reach the desired particle size. It has been found that this multistage growth is advantageous since the polymerization process conditions can be separately optimized for each growth stage, and it allows the final growth stage to be effected using process conditions and controls conventionally used in the suspension polymerization production of millimeter-sized particles.

The amine-functionalized particles may be reacted further to couple additional chemical functions to the amine groups or to transform the amine groups into other nitrogen attached functional groups. An example of the preparation is as follows (162):

Preparation 3–16: A reactor was charged with 1929 *kg* of an aqueous suspension of 55 *kg* 20 *μ.m* PS Dynospheres®, 18 *kg* of cellulose ether (Methocel K100, pre-dissolved in water) and 1600 *kg* water. The suspension was stirred at 40 rpm and heated to 40°C over 0.5 *h*.

Next, 1.0 *kg* dibenzoyl peroxide (75% in water) was dissolved in 10 *kg* styrene in a 10 *L* vessel using a conventional propeller as an agitator. After complete dissolution this was charged to the reactor. The suspension was kept at 40°C for 1 *h* then raised to 80°C over 1.5 *h*.

A styrene monomer emulsion was prepared by mixing 385 *kg* styrene, 3.0 *kg* dibenzoyl peroxide (75% in water) for 30 *min*. Then 770 *kg* water and 1.66 *kg* Tween 20 stabilizer were added and the mixture was emulsified and added to the reactor over 8 *h* at rates of 90.75 $kg\,h^{-1}$, 115.09 $kg\,h^{-1}$, 133.5 $kg\,h^{-1}$, 146.3 $kg\,h^{-1}$, 156.2 $kg\,h^{-1}$, 165 $kg\,h^{-1}$, 173 $kg\,h^{-1}$ and 177.5 $kg\,h^{-1}$ for 1 *h* each.

After 10 *min* at 80°C, the reactor was charged with 5 *g* polymerization inhibitor KI dissolved in 12.5 *g* water and after 2 *h* at 80°C a further 15 *g* KI dissolved in 37.5 *g* water was added.

After the polymerization reaction was complete, the reaction mixture was held at 80°C for a further 2 *h*.

The product was recovered and analyzed for particle size distribution using a Coulter Counter 256. Mode diameter: 39 41 *μm* CV: 5 6%.

Also, the fabrication of porous crosslinked PS particles containing an amine functionality have been described (163):

Preparation 3–17: In a two-stage Manton-Gaulin homogenizer, 850 *g* of water, 110.50 *g* of bis(2-ethylhexyl)adipate, 141.95 *g* of acetone and 4.25 *g* of sodium dodecyl sulfate were homogenized at 400 $kg\,cm^{-2}$ in the first stage and 100 $kg\,cm^{-2}$ in the second stage for 8–9 *min*.

After homogenization, 102.68 *g* of the emulsion was charged with a seed suspension of monodisperse oligomeric styrene particles having a particle diameter of 5 *μm*; 27.21 *g* of seed suspension containing 1.71 *g* of oligomeric particles and 26.2 *g* of water was used.

After stirring at 45°C for 24 *h*, 87.06 *g* of the seed suspension containing activated seed particles were charged to 1436.08 *g* of an emulsion containing 1035.84 *g* of water, 1.58 *g* of Methocel K100, 0.5 *g* of sodium dodecyl sulfate, 53.41 *g* of 80% divinylbenzene [i.e. 80% by weight DVB, 20% by weight ethyl vinyl benzene and other by-products in DVB production], 56.07 *g* of styrene, 6.71 *g* of 2,2′-azobis(2-methylbutyronitrile), c.f. Figure 3.74, 269.41 *g* of toluene and 12.56 *g* of 4-aminostyrene. The emulsion was homogenized without addition of 4-aminostyrene at 400 $kg\,cm^{-2}$ in the first stage and 100 $kg\,cm^{-2}$ in the second stage for 8–9 *min* before the emulsion was mixed with 4-aminostyrene.

After swelling at 27°C for 1 *h*, a mixture of 473.69 *g* of water and 3.16 *g* of Methocel K100 was then charged to the reactor. The dispersion was then polymerized for 1 *h* at 60°C and 10 *h* at 70°C, yielding a suspension of particles having a diameter of 30 *μm*.

The particles were separated from the liquid phase by flotation and the liquid phase was discharged. The particles were then cleaned with 2 *l* of methanol by stirring for 1 *h* followed by sedimentation. After sedimentation the liquid phase was discharged, new methanol (2 *l*) was charged and the described procedure was repeated 4 times. The particle suspension was then sieved through a 100 *μm* sieving cloth. Then the particle suspension was diafiltered with 6 *l* of butylacetate followed by 6.7 *l* of methanol. Finally, the particles were cleaned by sedimentation and discharging of the liquid phase, with 2 *l* of methanol minimum 3 times.

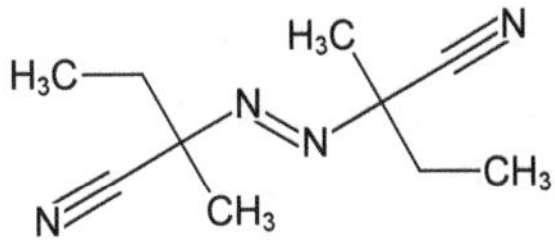

Figure 3.74 2,2′-Azobis(2-methylbutyronitrile).

Also, a further functionalization of amine-functionalized particles with carboxyl and amide functionalities has been detailed (163).

3.20 Nanomaterials

3.20.1 Dopamine-Functionalized Multiwalled Carbon Nanotubes

Considerable attention has been payed to CNTs as a new class of nanomaterials. Both single-walled CNTs and multiwalled CNTs have several interesting properties such as high-aspect-ratio, ultra-light weight, tremendous strength, high thermal conductivity and remarkable electronic properties ranging from metallic to semiconducting (164).

The chemical bonding of CNT is composed entirely of sp^2 carbon-carbon bonds. This bonding structure, stronger than the sp^3 bonds found in diamond, results in CNT materials with extremely high mechanical properties (165).

Compared with traditional reinforcing agents such as glass fibers, CNTs are much more efficient in improving the composite properties because of their extremely high aspect ratio. However, achieving a high degree of dispersion of CNTs in a polymer matrix is challenging due to agglomeration and aggregation into bundles (166–168).

In order to solve these problems, several strategies for synthesis of the nanotube-reinforced polymer matrix composites have been developed, including solution-casting with ultrasonication, melt-mixing, surfactant-assisted processing, surface modification strategies, e.g., grafting and polymer wrapping, and *in-situ* polymerization of monomers in the presence of CNTs (169).

Also, many approaches have been tried to carry out a chemical functionalization of carbon nanotubes. For example, the generation of surface hydrophilic substituents, such as carboxylic, hydroxyl or sulfonic acid groups, by suitable chemical method is rather easy for their wide use in medical and biological applications (170–172).

The design and synthesis of copolymers with the introduction of flexible linkages, such as amide, ester, ether, and sulfide linkages between the aromatic rings of the main chain, is a successful way to create more tractable polymers. Poly(amide-imide)s that have amide and imide linkages in the main backbone are a class of high-performance materials which possess desirable characteristics with the merits of both poly(amide)s and poly(imide)s, for example, exceptional thermomechanical properties at high temperature with good dimensional stability, excellent creep, and chemical resistance.

These materials have found extensive use in many engineering applications, such as molded parts for the space shuttle, engine parts of world-class racing cars, and many other critical components like membranes for ultrafiltration (170, 173).

An easy and feasible method developed to functionalize multiwalled CNTs with dopamine biomolecule in a one-pot procedure has been described (169).

The model poly(amide imide) used in this study was an amino acid-containing polymer with a pendant dopamine moiety. The introduction of several functional groups as well as dopamine and amino acid bulky substituents will result in increasing chain packing distances and decrease intermolecular interactions, leading to better interaction of the poly(amide imide) chains with multiwalled CNTs and a better dispersion in the polymer matrix. In addition, because of the dopamine-functionalization of multiwalled CNTs, a better compatibility and interaction between the multiwalled CNTs and the polymer could be found (169).

As basic monomers, 3,5-diamino-*N*-(3,4-dihydroxyphenethyl)-benzamide and *N*,*N'*-(pyromellitoyl)-bis-*L*-isoleucine as diamine monomers were used. These compounds are shown in Figure 3.75.

3,5-Diamino-*N*-(3,4-dihydroxyphenethyl)benzamide and *N*,*N'*-(pyromellitoyl)-bis-*L*-isoleucine were copolymerized under microwave radiation at 120°C for 4 *min*. Then the polymers were attached to the functionalized multiwalled CNTs, which were previously functionalized with carboxylic groups or dopamine.

3.20.2 *Polymer Composites with Functionalized Nanoparticles*

The state-of-the-art of polymer composites with functionalized nanoparticles has been discussed in a monograph (174).

Functionalization can be defined as an addition of a chemical or a physical functionality on the surface of the nanoparticles. Some modification methods involve the direct functionalization with a functional ligand containing one reactive group able to attach onto the filler surface, and the other representing the required active surface functionality.

Direct functionalization allows the modification of the surface of nanoparticles in one single step. However, in some cases a competitive reaction with functionality, which is going to be introduced on

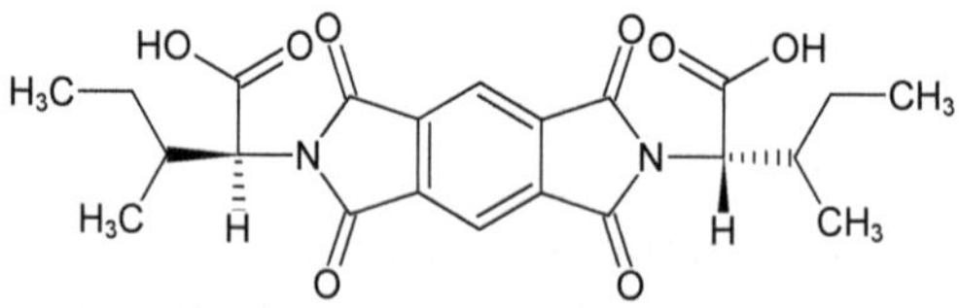

N,N'-(Pyromellitoyl)-bis-*L*-isoleucine

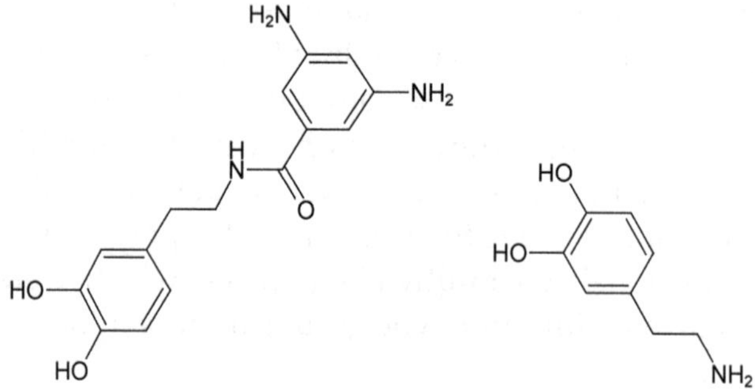

3,5-Diamino-*N*-(3,4-dihydroxyphenethyl)benzamide Dopamine

Figure 3.75 Diamine monomers.

the surface of the nanofillers, can occur, and thus a two-step modification needs to be applied. Postfunctionalization via thiol-ene reactions, Diels-Alder reactions, or Huisgen reactions can also be applied to control the surface chemistry and energy (174).

3.20.3 Antifouling Enhancement of Nanocomposite Membranes

Carboxylic acid and sulfate functional groups were coated on the surface of zirconia nanoparticles. The functionalized nanoparticles were embedded in the matrix of poly(acrylonitrile) and PSf membranes, and nanocomposite membrane models were synthesized (175).

The aim of the study was to exploit the proper arrangement of nanoparticles in the membrane matrix to strengthen the antifouling properties of the membrane. The change made in the superficial and structural properties of the membrane using functional groups caused internal pore blocking.

3.21 Sensitive Detection of Explosives

Functional poly(dihalopentadiene)s were synthesized by boron trihalide-mediated multicomponent polymerization routes in a stereoselective manner (176). The polymerization of tetraphenylethylene-containing diyne, BX3 (X = Cl, Br) and *p*-tolualdehyde (4-methylbenzaldehyde) proceeds smoothly in dichloromethane under mild conditions to afford high molecular weight poly(dihalopentadiene)s with a predominant (Z,Z)-configuration in moderate to good yields. A model compound used for the preparation is shown in Figure 3.76.

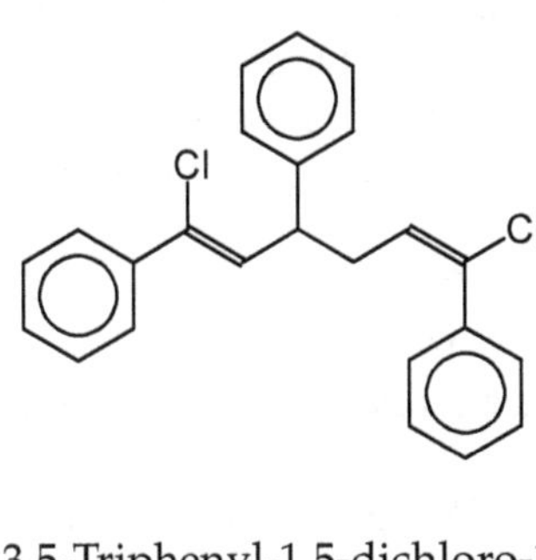

Figure 3.76 1,3,5-Triphenyl-1,5-dichloro-1,4-pentadiene.

The reaction conditions and the boron trihalide used were found to have great effects on the stereochemistry of the resulting polymer structures. The obtained poly(1,5-dihalo-(Z,Z)-1,4-pentadiene)s show a high thermal stability and good film-forming ability. Their thin films show high refractive index of 1.9007-1.6462 in a wide wavelength region of 380-890 *nm* with low optical dispersion. The polymers are weakly emissive in dilute solutions but become highly emissive upon aggregation, demonstrating a unique phenomenon of aggregation-enhanced emission.

Their nanoaggregates in aqueous media can serve as sensitive fluorescent chemosensors for the detection of explosives with a superamplification effect and a low detection limit (176). To demonstrate the potential of the polymers for explosive detection, picric acid, c.f. Figure 3.77, was selected as a model analyte.

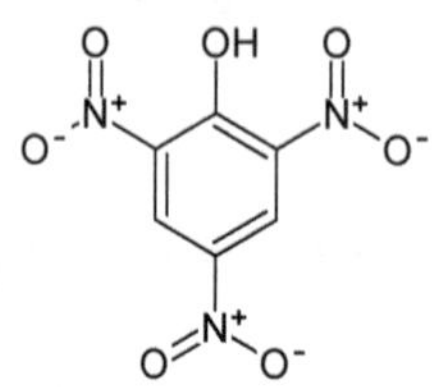

Figure 3.77 Picric acid.

References

1. K.R. Reddy, B. Hemavathi, G.R. Balakrishna, A. Raghu, S. Naveen, and M. Shankar, *Organic conjugated polymer-based functional nanohybrids: Synthesis methods, mechanisms and its applications in electrochemical energy storage supercapacitors and solar cells* in K. Pielichowski and T.M. Majka, eds., *Polymer Composites with Functionalized Nanoparticles*, Micro and Nano Technologies, chapter 11, pp. 357–379. Elsevier, 2019.
2. C. Yu, H. Li, J. Luo, M. Zheng, W. Zhong, and W. Yang, *Electrochimica Acta*, Vol. 284, p. 69, 2018.
3. T.M.S. David, C. Zhang, and S.-S. Sun, *Journal of Chemistry*, Vol. 2014, 2014.
4. J. Zhang, C. Yan, W. Wang, Y. Xiao, X. Lu, S. Barlow, T.C. Parker, X. Zhan, and S.R. Marder, *Chemistry of Materials*, Vol. 30, p. 309, 2018.

5. J.-H. Cho, C.-S. Sone, D.-Y. Kim, H.-G. Hong, and S.-S. Kim, Method of forming fine pattern using azobenzene-functionalized polymer and method of manufacturing nitride-based semiconductor light emitting device using the method of forming fine pattern, US Patent 7 943 290, assigned to Samsung LED Co., Ltd. (KR), May 17, 2011.
6. S.-S. Kim, J. Jo, C. Chun, J.-C. Hong, and D.-Y. Kim, *Journal of Photochemistry and Photobiology A: Chemistry*, Vol. 188, p. 364 , 2007.
7. B.K. Mandal, R.J. Jeng, J. Kumar, and S.K. Tripathy, *Die Makromolekulare Chemie, Rapid Communications*, Vol. 12, p. 607, 1991.
8. C. Yao, X. Yin, Y. Yu, Z. Cai, and X. Wang, *Advanced Functional Materials*, Vol. 27, p. 1700794, 2017.
9. G. Lv, S. Zhang, G. Wang, J. Shao, H. Tian, and D. Yu, *Reactive and Functional Polymers*, Vol. 111, p. 44, 2017.
10. Z.-A. Lan, Y. Fang, Y. Zhang, and X. Wang, *Angewandte Chemie*, Vol. 130, p. 479, 2018.
11. L. Wang, Y. Zhang, L. Chen, H. Xu, and Y. Xiong, *Advanced Materials*, Vol. 0, p. 1801955, 2018. in press.
12. S. Yanagida, A. Kabumoto, K. Mizumoto, C. Pac, and K. Yoshino, *J. Chem. Soc., Chem. Commun.*, pp. 474–475, 1985.
13. X. Wang, K. Maeda, A. Thomas, K. Takanabe, G. Xin, J.M. Carlsson, K. Domen, and M. Antonietti, *Nature Materials*, Vol. 8, p. 76, November 2008.
14. P. Guiglion, C. Butchosa, and M.A. Zwijnenburg, *Macromolecular Chemistry and Physics*, Vol. 217, p. 344, 2015.
15. V.S. Vyas, V.W.-H. Lau, and B.V. Lotsch, *Chemistry of Materials*, Vol. 28, p. 5191, 2016.
16. D. Liu, J. Wang, X. Bai, R. Zong, and Y. Zhu, *Advanced Materials*, Vol. 28, p. 7284, 2016.
17. L. Wang, Y. Wan, Y. Ding, Y. Niu, Y. Xiong, X. Wu, and H. Xu, *Nanoscale*, Vol. 9, p. 4090, 2017.
18. L. Wang, X. Zheng, L. Chen, Y. Xiong, and H. Xu, *Angewandte Chemie International Edition*, Vol. 57, p. 3454, 2018.
19. J.S. Paneysar, S. Barton, S. Chandra, P. Ambre, and E. Coutinho, *Water Science and Technology*, Vol. 75, p. 1084, 2016.
20. A. Bhatnagar and M. Sillanpää, *Advances in Colloid and Interface Science*, Vol. 152, p. 26, 2009.
21. D.W. Kang, M. Kang, M. Moon, H. Kim, S. Eom, J.H. Choe, W.R. Lee, and C.S. Hong, *Chemical Science*, Vol. 9, p. 6871, 2018.
22. Y. Lu, Z. Wu, M. Li, Q. Liu, and D. Wang, *Reactive and Functional Polymers*, Vol. 82, p. 98, 2014.
23. L. Trzonkowska, B. Leśniewska, and B. Godlewska-Żyłkiewicz, *Reactive and Functional Polymers*, Vol. 117, p. 131, 2017.
24. M. Cegłowski, B. Gierczyk, M. Frankowski, and Ł. Popenda, *Reactive and Functional Polymers*, Vol. 131, p. 64 , 2018.

25. N. Saman, M.U. Rashid, J.W.P. Lye, and H. Mat, *Reactive and Functional Polymers*, Vol. 123, p. 106, 2018.
26. F.-Q. An, R.-Y. Wu, M. Li, T.-P. Hu, J.-F. Gao, and Z.-G. Yuan, *Reactive and Functional Polymers*, Vol. 118, p. 42, 2017.
27. B.A. Omondi, H. Okabe, Y. Hidaka, and K. Hara, *Reactive and Functional Polymers*, Vol. 130, p. 90, 2018.
28. Q. Li, B. Ling, L. Jiang, and L. Ye, *Chemical Engineering Journal*, Vol. 350, p. 217, 2018.
29. T. Kitao, Y. Zhang, S. Kitagawa, B. Wang, and T. Uemura, *Chem. Soc. Rev.*, Vol. 46, p. 3108, 2017.
30. K.J. Lee, J.H. Lee, S. Jeoung, and H.R. Moon, *Accounts of Chemical Research*, Vol. 50, p. 2684, 2017.
31. D.V. Amato, H. Lee, J.G. Werner, D.A. Weitz, and D.L. Patton, *ACS Applied Materials & Interfaces*, Vol. 9, p. 3288, 2017.
32. H. Lee, C.-H. Choi, A. Abbaspourrad, C. Wesner, M. Caggioni, T. Zhu, S. Nawar, and D.A. Weitz, *Advanced Materials*, Vol. 28, p. 8425, 2016.
33. House Committee on Energy and Commerce, Microbead-free waters act of 2015, Public Law H.R. 1321, 2015.
34. H.J. Jeong and B.K. Kim, *Reactive and Functional Polymers*, Vol. 116, p. 92, 2017.
35. P. Liu, Organic acid- or latent organic acid-functionalized polymer-coated metal powders for solder pastes, US Patent 9 682 447, assigned to Henkel IP & Holding GmbH (Düsseldorf, DE), June 20, 2017.
36. L.-F. Wang and J.-W. Rhim, *International Journal of Biological Macromolecules*, Vol. 80, p. 460, 2015.
37. Z. Wu, J. Wu, T. Peng, Y. Li, D. Lin, B. Xing, C. Li, Y. Yang, L. Yang, L. Zhang, R. Ma, W. Wu, X. Lv, J. Dai, and G. Han, *Polymers*, Vol. 9, 2017.
38. S.-H. Liao, P.-L. Liu, M.-C. Hsiao, C.-C. Teng, C.-A. Wang, M.-D. Ger, and C.-L. Chiang, *Industrial & Engineering Chemistry Research*, Vol. 51, p. 4573, 2012.
39. W. Hu, L. Song, L. Wang, Y. Hu, and P. Zhang, *Fire Safety Science*, Vol. 11, p. 895, 2014.
40. X. Wang, W. Xing, X. Feng, B. Yu, L. Song, and Y. Hu, *Polym. Chem.*, Vol. 5, p. 1145, 2014.
41. Q. He, T. Yuan, X. Yan, D. Ding, Q. Wang, Z. Luo, T.D. Shen, S. Wei, D. Cao, and Z. Guo, *Macromolecular Chemistry and Physics*, Vol. 215, p. 327, 2014.
42. S.-D. Jiang, G. Tang, Z.-M. Bai, Y.-Y. Wang, Y. Hu, and L. Song, *RSC Adv.*, Vol. 4, p. 3253, 2014.
43. G. Huang, Z. Fei, X. Chen, F. Qiu, X. Wang, and J. Gao, *Applied Surface Science*, Vol. 258, p. 1011510122, 2012.
44. Y. Bi, X.N. Zhang, Y. Jiang, and S.L. Sun, *Advanced Materials Research*, Vol. 1053, p. 201, 12 2014.

45. W.J.J. Op de Beeck, L.E.D. Deprez, and G.G.P. Deroover, Liquid toner dispersion and use thereof, US Patent 9 482 979, assigned to Xeikon IP BV (Eede, NL), November 1, 2016.
46. J.A.M. van Broekhoven, Hydrogenation of CO/olefin copolymer, US Patent 4 929 701, assigned to Shell Oil Company (Houston, TX), May 29, 1990.
47. P.K. Wong, Catalytic reduction of carbon monoxide/olefin copolymer to polyalcohol, US Patent 4 868 254, assigned to Shell Oil Company (Houston, TX), September 19, 1989.
48. G.G. Hlatky, Reduction of α-olefin-carbon monoxide copolymers to polyalcohols with borohydride salt, US Patent 5 300 596, assigned to Gregory G. Hlatky, April 5, 1994.
49. G.H. Hofmann and S.C.F. Lee, Hydroxyl-functionalized polymer compositions, US Patent 6 534 602, assigned to E. I. du Pont de Nemours and Company (Wilmington, DE), March 18, 2003.
50. Y. Zhao, Y. Fu, Y. He, B. Hu, L. Liu, J. Lü, and C. Lü, *RSC Adv.*, Vol. 5, p. 93480, 2015.
51. B. Bae, T. Yoda, K. Miyatake, H. Uchida, and M. Watanabe, *Angewandte Chemie International Edition*, Vol. 49, p. 317, 2009.
52. J.J. Kaczur, H. Yang, S.D. Sajjad, Z. Liu, and R.I. Masel, *Frontiers in Chemistry*, Vol. 6, p. 263, 2018.
53. R.L. Masel, Q. Chen, Z. Liu, and R. Kutz, Ion-conducting membranes, US Patent 9 370 773, assigned to Dioxide Materials, Inc. (Boca Raton, FL), June 21, 2016.
54. R.I. Masel, Q. Chen, Z. Liu, and R. Kutz, Ion-conducting membranes, US Patent 9 580 824, assigned to Dioxide Materials, Inc. (Boca Raton, FL), February 28, 2017.
55. R.I. Masel, S.D. Sajjad, Y. Gao, Z. Liu, and Q. Chen, Ion-conducting membranes, US Patent 9 849 450, assigned to Dioxide Materials, Inc. (Boca Raton, FL), December 26, 2017.
56. M. Pellerite, M. Kaplun, C. Hartmann-Thompson, K.A. Lewinski, N. Kunz, T. Gregar, J. Baetzold, D. Lutz, M. Quast, Z. Liu, H. Yang, S.D. Sajjad, Y. Gaob, and R. Masel, *ECS Transactions*, Vol. 80, p. 945, 2017.
57. J. Gao, S. Japip, and T.-S. Chung, *Chemical Engineering Journal*, Vol. 353, p. 689 , 2018.
58. B. Kanjilal, I. Noshadi, J.R. McCutcheon, A.D. Asandei, and R.S. Parnas, *Journal of Membrane Science*, Vol. 486, p. 59, 2015.
59. N.V. Blinova and F. Svec, *J. Mater. Chem. A*, Vol. 2, p. 600, 2014.
60. Y.S. Park, C. Ha, and S.W. Kang, *RSC Advances*, Vol. 7, p. 33568, 2017.
61. D. Aaron and C. Tsouris, *Separation Science and Technology*, Vol. 40, p. 321, 2005.
62. H. Yang, Z. Xu, M. Fan, R. Gupta, R.B. Slimane, A.E. Bland, and I. Wright, *Journal of Environmental Sciences*, Vol. 20, p. 14, 2008.

63. N. Du, H.B. Park, M.M. Dal-Cin, and M.D. Guiver, *Energy Environ. Sci.*, Vol. 5, p. 7306, 2012.
64. J. Huang, J. Zou, and W.S.W. Ho, *Industrial & Engineering Chemistry Research*, Vol. 47, p. 1261, 2008.
65. N. Du, H.B. Park, G.P. Robertson, M.M. Dal-Cin, T. Visser, L. Scoles, and M.D. Guiver, *Nature Materials*, Vol. 10, p. 372, April 2011.
66. N.B. Mckeown, P.M. Budd, K. Msayib, and B. Ghanem, Microporous polymer material, WO Patent 2 005 012 397, assigned to The University of Manchester, February 10, 2005.
67. N.B. McKeown, P.M. Budd, and D. Fritsch, Thin layer composite membrane, WO Patent 2 005 113 121, assigned to The University of Manchester, GKSS Forschungszentrum Geesthact GmbH, December 01, 2005.
68. C. Liu, M.E. Schott, and T.C. Bowen, Tetrazole functionalized polymer membranes, US Patent 8 814 982, assigned to UOP LLC (Des Plaines, IL), August 26, 2014.
69. P.M. Budd, N.B. McKeown, and D. Fritsch, *Macromolecular Symposia*, Vol. 245-246, p. 403, 2007.
70. N. Jusoh, K.K. Lau, A.M. Shariff, and Y.F. Yeong, *International Journal of Greenhouse Gas Control*, Vol. 22, p. 213, 2014.
71. M.-B. Hägg, *Gas permeation unit (GPU)* in E. Drioli and L. Giorno, eds., *Encyclopedia of Membranes*, pp. 849–849. Springer Berlin Heidelberg, Berlin, Heidelberg, 2016.
72. T. Sandu, A. Sarbu, C.M. Damian, D. Patroi, T.V. Iordache, T. Budinova, B. Tsyntsarski, M.F. Yardim, and A. Sirkecioglu, *Reactive and Functional Polymers*, Vol. 96, p. 5, 2015.
73. T. Srisook, T. Vongsetskul, J. Sucharitakul, P. Chaiyen, and P. Tangboriboonrat, *Reactive and Functional Polymers*, Vol. 82, p. 41, 2014.
74. C.D. Sanborn and S. Ardo, Photoelectrochemical ion pumping with dye-functionalized polymer membranes, in *Meeting Abstracts*, number 37, pp. 2011–2011. The Electrochemical Society, 2015.
75. X. Lin, K. Wang, Y. Feng, J.Z. Liu, X. Fang, T. Xu, and H. Wang, *Journal of Membrane Science*, Vol. 482, p. 67, 2015.
76. A. Farrukh, F. Ashraf, A. Kaltbeitzel, X. Ling, M. Wagner, H. Duran, A. Ghaffar, H. ur Rehman, S.H. Parekh, K.F. Domke, et al., *Polymer Chemistry*, Vol. 6, p. 5782, 2015.
77. R.P. Pandey, A.K. Thakur, and V.K. Shahi, *ACS Applied Materials & Interfaces*, Vol. 6, p. 16993, 2014.
78. L. Cao, L. Kong, L. Kong, X. Zhang, and H. Shi, *Journal of Power Sources*, Vol. 299, p. 25, 2015.
79. Z. Wei, Y. Liu, H. Hu, J. Yu, and F. Li, *RSC Adv.*, Vol. 6, p. 108240, 2016.
80. Y. Araki, Y. Kobayashi, T. Kawaguchi, T. Kaneko, and N. Arai, *Journal of Membrane Science*, Vol. 564, p. 184, 2018.

81. M. Peng, H. Li, R. Long, S. Shi, H. Zhou, and S. Yang, *Molecules*, Vol. 23, p. 1554, June 2018.
82. W.J. Koros and R.L. Burns, Dithiolene functionalized polymer membrane for olefin/paraffin separation, US Patent 7 160 356, assigned to Board of Regents, The University of Texas System (Austin, TX), January 9, 2007.
83. R.A. Hayes, Polyimide gas separation membranes, US Patent 4 717 394, assigned to E. I. Du Pont de Nemours and Company (Wilmington, DE), January 5, 1988.
84. H.H. Kung, J. Shen, M.C. Kung, and C.M. Hayner, Polymer functionalized graphene oxide and thermally responsive ion permeable membranes made therefrom, US Patent Application 20 150 318 531, assigned to Northwestern University, November 5, 2015.
85. R. Knoesel, M. Ehrmann, and J. Galin, *Polymer*, Vol. 34, p. 1925, 1993.
86. J.G. Weers, J.F. Rathman, F.U. Axe, C.A. Crichlow, L.D. Foland, D.R. Scheuing, R.J. Wiersema, and A.G. Zielske, *Langmuir*, Vol. 7, p. 854, 1991.
87. D. Han, P. Xiao, J. Gu, J. Chen, Z. Cai, J. Zhang, W. Wang, and T. Chen, *RSC Adv.*, Vol. 4, p. 22759, 2014.
88. L. Ma, Functionalized polymer, rubber composition and pneumatic tire, US Patent 9 574 042, assigned to The Goodyear Tire & Rubber Company (Akron, OH), February 21, 2017.
89. L. Ma and M.M. Vielhaber, Functionalized polymer, rubber composition, and pneumatic tire, US Patent 9 790 289, assigned to The Goodyear Tire & Rubber Company (Akron, OH), October 17, 2017.
90. D. Rivin, J. Aron, and A.I. Medalia, *Rubber Chemistry and Technology*, Vol. 41, p. 330, 1968.
91. A.M. Gessler, W.M. Hess, and A.I. Medalia, *Plastics & Rubber: Processing*, p. 141, 1978.
92. S.-H. Choi, M.-S. Kim, C.-J. Kim, and W.-M. Choi, Aminosilane terminal modifier to which functional group has been introduced, method for producing terminal-modified conjugated diene polymer using the aminosilane terminal modifier, and terminal-modified conjugated diene polymer produced according to the method, US Patent 9 951 150, assigned to LG Chem, Ltd. (KR), April 24, 2018.
93. Z. Qin, T.E. Hogan, T. Uchiyama, J.P. Abell, and S. Luo, Functionalized polymer, US Patent 9 556 297, assigned to Bridgestone Corporation (Tokyo, JP), January 31, 2017.
94. Y.-Y. Yan, Method of making a functionalized polymer with sulfide linkage, US Patent 9 260 540, assigned to Bridgestone Corporation (Tokyo, JP), February 16, 2016.
95. A. Hirao, S. Loykulnant, and T. Ishizone, *Progress in Polymer Science*, Vol. 27, p. 1399, 2002.

96. T.E. Hogan, W. Hergenrother, Y.-Y. Yan, and D. Lawson, Use of sulfur containing initiators for anionic polymerization of monomers, US Patent 7 612 144, assigned to Bridgestone Corporation (JP), November 3, 2009.
97. D.R. Brumbaugh, S. Luo, C. Rademacher, and Y.-Y. Yan, Functionalized polymer, US Patent 8 389 644, assigned to Bridgestone Corp. (Tokyo, JP), March 5, 2013.
98. Y.-Y. Yan, Synthesis of a liquid polymer and a functionalized polymer, US Patent 7 981 990, assigned to Bridgestone Corporation (Tokyo, JP), July 19, 2011.
99. S. Luo, Polymers functionalized with polycyano compounds, US Patent 10 081 688, assigned to Bridgestone Corporation (Tokyo, JP), September 25, 2018.
100. S. Luo, Polymers functionalized with protected oxime compounds containing a cyano group, US Patent 10 030 083, assigned to Bridgestone Corporation (Tokyo, JP), July 24, 2018.
101. S. Luo, Polymers functionalized with unsaturated heterocycles containing a protected amino group, US Patent 9 469 706, assigned to Bridgestone Corporation (Tokyo, JP), October 18, 2016.
102. A.H. Tsou, D.S. Cushing, J.A. Mann, and A.O. Patil, Comb-block copolymers of isobutylene copolymer backbone with functional polymer comb arms, US Patent Application 20 180 291 133, assigned to ExxonMobil Chemical Patents Inc., October 11, 2018.
103. M.R. Sivik and R.A. Denis, Functionalized polymer composition for grease, US Patent 8 563 488, assigned to The Lubrizol Corporation (Wickliffe, OH), October 22, 2013.
104. M.R. Sivik and M.G. Fahmy, Functionalized polymer composition for grease, WO Patent 2 004 111 163, assigned to The Lubrizol Corporation (Wickliffe, OH), February 10, 2005.
105. ASTM International, Standard test method for determining the resistance of lubricating grease to water spray, ASTM Standard ASTM D4049-16, ASTM International, West Conshohocken, PA, 2016.
106. C.R. Scharf, S.R. Twining, and P.R. Todd, Functionalized polymer as grease additive, US Patent 6 300 288, assigned to The Lubrizol Corporation (Wickliffe, OH), October 9, 2001.
107. J. Fu, *Journal of Polymer Science Part B: Polymer Physics*, Vol. 56, p. 1336, 2018.
108. W. Zhao, X. Jin, Y. Cong, Y. Liu, and J. Fu, *Journal of Chemical Technology & Biotechnology*, Vol. 88, p. 327, 2012.
109. M. Mehrali, A. Thakur, C.P. Pennisi, S. Talebian, A. Arpanaei, M. Nikkhah, and A. Dolatshahi-Pirouz, *Advanced Materials*, Vol. 29, p. 1603612, 2016.
110. H. Yuk, T. Zhang, S. Lin, G.A. Parada, and X. Zhao, *Nature Materials*, Vol. 15, p. 190, November 2015.

111. L. Han, X. Lu, K. Liu, K. Wang, L. Fang, L.-T. Weng, H. Zhang, Y. Tang, F. Ren, C. Zhao, G. Sun, R. Liang, and Z. Li, *ACS Nano*, Vol. 11, p. 2561, 2017.
112. T. Shirakura, T.J. Kelson, A. Ray, A.E. Malyarenko, and R. Kopelman, *ACS Macro Letters*, Vol. 3, p. 602, 2014.
113. C. Ma, Y. Shi, D.A. Pena, L. Peng, and G. Yu, *Angewandte Chemie*, Vol. 127, p. 7484, 2015.
114. H. Gao, J.K. Hyun, M.H. Lee, J.-C. Yang, L.J. Lauhon, and T.W. Odom, *Nano Letters*, Vol. 10, p. 4111, 2010.
115. A. Richter, G. Paschew, S. Klatt, J. Lienig, K.-F. Arndt, and H.-J. Adler, *Sensors*, Vol. 8, p. 561, January 2008.
116. Y. Takashima, S. Hatanaka, M. Otsubo, M. Nakahata, T. Kakuta, A. Hashidzume, H. Yamaguchi, and A. Harada, *Nature Communications*, Vol. 3, p. 1270, December 2012.
117. X. Liu, B. He, Z. Wang, H. Tang, T. Su, and Q. Wang, *Scientific Reports*, Vol. 4, p. 6673, October 2014.
118. H. Yuk, S. Lin, C. Ma, M. Takaffoli, N.X. Fang, and X. Zhao, *Nature Communications*, Vol. 8, p. 14230, February 2017.
119. Y. Shi and G. Yu, *Chemistry of Materials*, Vol. 28, p. 2466, 2016.
120. Y. Shi, J. Zhang, L. Pan, Y. Shi, and G. Yu, *Nano Today*, Vol. 11, p. 738, 2016.
121. Y. Zhao, B. Liu, L. Pan, and G. Yu, *Energy Environ. Sci.*, Vol. 6, p. 2856, 2013.
122. Q. Chen, H. Chen, L. Zhu, and J. Zheng, *J. Mater. Chem. B*, Vol. 3, p. 3654, 2015.
123. J. Hou, X. Ren, S. Guan, L. Duan, G.H. Gao, Y. Kuai, and H. Zhang, *Soft Matter*, Vol. 13, p. 1357, 2017.
124. M.A. Haque, T. Kurokawa, and J.P. Gong, *Polymer*, Vol. 53, p. 1805, 2012.
125. T. Nonoyama, S. Wada, R. Kiyama, N. Kitamura, M.I. Mredha, X. Zhang, T. Kurokawa, T. Nakajima, Y. Takagi, K. Yasuda, and J.P. Gong, *Advanced Materials*, Vol. 28, p. 6740, 2016.
126. L. Carlsson, S. Rose, D. Hourdet, and A. Marcellan, *Soft Matter*, Vol. 6, p. 3619, 2010.
127. M. Takafuji, S.-Y. Yamada, and H. Ihara, *Chem. Commun.*, Vol. 47, p. 1024, 2011.
128. Q. Wang, R. Hou, Y. Cheng, and J. Fu, *Soft Matter*, Vol. 8, p. 6048, 2012.
129. J. Yang, X.-P. Wang, and X.-M. Xie, *Soft Matter*, Vol. 8, p. 1058, 2012.
130. P. Thoniyot, M.J. Tan, A.A. Karim, D.J. Young, and X.J. Loh, *Advanced Science*, Vol. 2, p. 1400010, 2015.
131. S. Rose, A. Prevoteau, P. Elzière, D. Hourdet, A. Marcellan, and L. Leibler, *Nature*, Vol. 505, p. 382, December 2013.

132. S. Rose, A. Dizeux, T. Narita, D. Hourdet, and A. Marcellan, *Macromolecules*, Vol. 46, p. 4095, 2013.
133. A. Vallés-Lluch, S. Poveda-Reyes, P. Amorós, D. Beltrán, and M. Monleón Pradas, *Biomacromolecules*, Vol. 14, p. 4217, 2013.
134. W.L. Murphy, T.C. McDevitt, and A.J. Engler, *Nature Materials*, Vol. 13, p. 547, May 2014.
135. J. Yang, L.-H. Deng, C.-R. Han, J.-F. Duan, M.-G. Ma, X.-M. Zhang, F. Xu, and R.-C. Sun, *Soft Matter*, Vol. 9, p. 1220, 2013.
136. S. Xia, S. Song, X. Ren, and G. Gao, *Soft Matter*, Vol. 13, p. 6059, 2017.
137. Y.-N. Sun, G.-R. Gao, G.-L. Du, Y.-J. Cheng, and J. Fu, *ACS Macro Letters*, Vol. 3, p. 496, 2014.
138. Y. Sun, S. Liu, G. Du, G. Gao, and J. Fu, *Chem. Commun.*, Vol. 51, p. 8512, 2015.
139. L. Li, R. Jiang, J. Chen, M. Wang, and X. Ge, *RSC Adv.*, Vol. 7, p. 1513, 2017.
140. P. Wang, G. Deng, L. Zhou, Z. Li, and Y. Chen, *ACS Macro Letters*, Vol. 6, p. 881, 2017.
141. T. Huang, H.-G. Xu, K. Jiao, L. Zhu, H. Brown, and H.-L. Wang, *Advanced Materials*, Vol. 19, p. 1622, 2007.
142. L. Xiao, J. Zhu, J.D. Londono, D.J. Pochan, and X. Jia, *Soft Matter*, Vol. 8, p. 10233, 2012.
143. J. Zhao, K. Jiao, J. Yang, C. He, and H. Wang, *Polymer*, Vol. 54, p. 1596, 2013.
144. T. Wang and Y. Luo, *International Journal of Molecular Sciences*, Vol. 19, p. 3112, October 2018.
145. X. Liu, J.C. Kim, A.L. Miller, B.E. Waletzki, and L. Lu, *New J. Chem.*, Vol. 42, p. 17671, 2018.
146. C. Liu, J. Han, Y. Pei, and J. Du, *Applied Sciences*, Vol. 8, p. 1941, October 2018.
147. B.A. Omondi, H. Okabe, Y. Hidaka, and K. Hara, *Reactive and Functional Polymers*, Vol. 125, p. 11, 2018.
148. K. Schöller, C. Toncelli, J. Experton, S. Widmer, D. Rentsch, A. Vetushka, C.J. Martin, M. Heuberger, C.E. Housecroft, E.C. Constable, L.F. Boesel, and L.J. Scherer, *RSC Adv.*, Vol. 6, p. 97921, 2016.
149. A. Akelah and D.C. Sherrington, *Chemical Reviews*, Vol. 81, p. 557, 1981.
150. B. Clapham, T.S. Reger, and K.D. Janda, *Tetrahedron*, Vol. 57, p. 4637, 2001.
151. C.C. Tzschucke, C. Markert, W. Bannwarth, S. Roller, A. Hebel, and R. Haag, *Angewandte Chemie International Edition*, Vol. 41, p. 3964, 2002.
152. B.L. Rivas, E. Pereira, and A. Maureira, *Polymer International*, Vol. 58, p. 1093, 2009.

153. N. Nishat, T. Ahamad, S. Ahmad, and S. Parveen, *Journal of Coordination Chemistry*, Vol. 64, p. 2639, 2011.
154. L. Timofeeva and N. Kleshcheva, *Applied Microbiology and Biotechnology*, Vol. 89, p. 475, February 2011.
155. A. Movahedi, K. Moth-Poulsen, J. Eklöf, M. Nydén, and N. Kann, *Reactive and Functional Polymers*, Vol. 82, p. 1, 2014.
156. P. Sund and C.-E. Wilén, *Reactive and Functional Polymers*, Vol. 89, p. 24, 2015.
157. A. Pourjavadi and A. Abedin-Moghanaki, *Reactive and Functional Polymers*, Vol. 105, p. 95, 2016.
158. Wikipedia contributors, Density functional theory — Wikipedia, the free encyclopedia, https://en.wikipedia.org/w/index.php?title=Density_functional_theory&oldid=849182898, 2018. [Online; accessed 9-October-2018].
159. P. Wang, Q. Ma, D. Hu, and L. Wang, *Reactive and Functional Polymers*, Vol. 91-92, p. 43, 2015.
160. M. Ruda and R. Slattegård, Polymer made of a primary amine functionalized polymer and a hemicellulose, US Patent 9 133 578, assigned to Cellutech AB (Stockholm, SE), September 15, 2015.
161. G. Fonnum, E. Weng, E.M. Aksnes, R. Nordal, P.C. Mork, S. Togersen, and J. Cockbain, Process for the preparation of functionalized polymer particles, US Patent 6 984 702, assigned to Dynal Biotech ASA (Oslo, NO), January 10, 2006.
162. J. Ugelstad and A. Berge, Process for the preparation of aqueous dispersions of organic material and possible further conversion to a polymer dispersion when the organic material is a polymerizable monomer, US Patent 4 530 956, assigned to Sintef (Trondheim NTH, NO), July 23, 1985.
163. A. Jorgedal, E. Aksnes, G. Fonnum, A. Molteberg, R. Nordal, H. Pettersen, T. Taarneby, S. Staale, E. Weng, F. Hansen, S. Nordbo, O. Aune, A. Berge, J. Bjorgum, T. Ellingsen, and J. Ugelstad, Process for the preparation of monodisperse polymer particles, US Patent 8 658 733, assigned to Life Technologies AS (Oslo, NO), February 25, 2014.
164. A. Agarwal, S.R. Bakshi, and D. Lahiri, *Carbon Nanotubes: Reinforced Metal Matrix Composites*, CRC Press, 2016.
165. N.G. Sahoo, S. Rana, J.W. Cho, L. Li, and S.H. Chan, *Progress in Polymer Science*, Vol. 35, p. 837.867, 2010.
166. B. Ruelle, S. Peeterbroeck, C. Bittencourt, G. Gorrasi, G. Patimo, M. Hecq, R. Snyders, S.D. Pasquale, and P. Dubois, *Reactive and Functional Polymers*, Vol. 72, p. 383, 2012.
167. Q. Cheng, M. Li, L. Jiang, and Z. Tang, *Advanced Materials*, Vol. 24, p. 1838, 2012.
168. W. Cui, M. Li, J. Liu, B. Wang, C. Zhang, L. Jiang, and Q. Cheng, *ACS Nano*, Vol. 8, p. 9511, 2014.

169. S. Mallakpour and A. Zadehnazari, *Reactive and Functional Polymers*, Vol. 106, p. 112 , 2016.
170. Z. Zhao, Z. Yang, Y. Hu, J. Li, and X. Fan, *Applied Surface Science*, Vol. 276, p. 476, 2013.
171. S. Mallakpour and A. Zadehnazari, *Progress in Organic Coatings*, Vol. 77, p. 679, 2014.
172. B.I. Kharisov, O.V. Kharissova, U.O. Méndez, and I.G.D.L. Fuente, *Synthesis and Reactivity in Inorganic, Metal-Organic, and Nano-Metal Chemistry*, Vol. 46, p. 55, 2016.
173. Y. Imai, *Reactive and Functional Polymers*, Vol. 30, p. 3, 1996. International Symposium on Condensation Polymers.
174. K. Pielichowski and T.M. Majka, eds. Micro and Nano Technologies. Elsevier, 2019.
175. K. Monsef, M. Homayoonfal, and F. Davar, *Reactive and Functional Polymers*, Vol. 131, p. 299, 2018.
176. T. Han, Y. Zhang, B. He, J. Lam, and B. Tang, *Polymers*, Vol. 10, p. 821, July 2018.

4

Medical Applications

The modification of polymers for controlled drug release has been reviewed (1). Incorporating labile bonds inside a polymer backbone and side chains yields interesting polymer materials that are responsive to the change of environmental stimuli. Drugs can be conjugated to various polymers through different conjugation linkages and spacers.

One of the key factors influencing the release profile of conjugated drugs is the hydrolytic stability of the conjugated linkage. Generally, the hydrolysis of acid-labile linkages, including acetal, imine, hydrazone, and to some extent β-thiopropionate, are relatively fast and the conjugated drug can be completely released in the range of several hours to a few days. The cleavage of ester linkages are usually slow, which is beneficial for a continuous and prolonged release.

Another key structural factor is the water solubility of polymer-drug conjugates. Generally, the release rate from highly water-soluble prodrugs is fast. In prodrugs with large hydrophobic segments, hydrophobic drugs are usually located in the hydrophobic core of the micelles and nanoparticles, which limits the access to the water, hence significantly lowering the hydrolysis rate. Finally, self-immolative polymers are also an intriguing new class of materials (1).

4.1 Biomedical Applications

Biodegradable polymers have been widely used in biomedical fields because of their biocompatibility and biodegradability (2). The development of biotechnology and medical technology has set higher requirements for biomedical materials. Novel biodegradable polymers with specific properties are a great demand.

Biodegradable polymers can be classified as natural or synthetic polymers according to their source. Synthetic biodegradable polymers have found more versatile and diverse biomedical applications owing to their tailorable designs or modifications.

Various types of synthetic biodegradable polymers with reactive groups and bioactive groups have been reviewed (2). Also, their structure, preparation procedures and properties were described.

Aliphatic polyesters with reactive groups are of interest, because of the demand for synthetic biopolymers with tunable properties, including features such as hydrophilicity, biodegradation rates, bioadhesion, and drug/targeting moiety attachment (3). In particular, polymeric biomaterials with properties that can be tailored by introducing functional groups, such as carboxyl, hydroxyl, amino, ketal, bromo, chloro, carbon-carbon double bonds or triple bonds, etc., are needed (2).

Aliphatic polyesters containing reactive groups can be prepared by the homopolymerization or copolymerization of cyclic monomers bearing protected functional groups.

4.1.1 Live Cell Surfaces

The capability to graft synthetic polymers onto the surfaces of live cells offers the potential to manipulate and control their phenotype and underlying cellular processes (4). However, conventional grafting strategies for conjugating preformed polymers to cell surfaces are limited by a low polymer grafting efficiency.

An alternative grafting-from strategy has been reported for directly engineering the surfaces of live yeast and mammalian cells through cell surface-initiated controlled radical polymerization (4). By the development of a cytocompatible photoinduced electron transfer-reversible addition-fragmentation chain transfer polymerization, synthetic polymers with a narrow polydispersity of $M_w/M_n < 1.3$ could be obtained at room temperature in 5 *min*. This polymerization strategy enables the chain growth to be initiated directly from chain transfer agents that are anchored on the surface insertion used, while maintaining a high cell viability.

In comparison to conventional grafting-to approaches, these methods can significantly improve the efficiency of grafting polymer chains and enable the active manipulation of cellular phenotypes (4).

4.1.2 *Functional Fibers for Biomedical Applications*

4.1.2.1 *Spinning Methods*

The advantage of the electrospinning fabrication process lies in its simplicity, cost-effectiveness, and the possibility for scaled-up fabrication (5). Usually, electrospinning takes place at room temperature in a dry environment. Microfluidic spinning is typically a wet spinning process. Microfluidic spinning can continuously produce microfibers with a uniform diameter and spatiotemporal control. A review has been presented that compares these two methods (5).

The modification of polymers with functional groups prior to electrospinning offers the opportunity to control the spatial presentation of functional groups within a scaffold, as well as to incorporate multiple bioactive cues. The methods to modify poly(caprolactone) (PCL) with peptides and electrospinning these peptide-PCL conjugates to functionalize a scaffold surface in a single step have been detailed in a monograph (6). Also, methods to adapt standard electrospinning setups to create single-peptide or dual-peptide gradients within a single construct have been described.

In electrospinning, many biocompatible and biodegradable synthetic and natural polymers can be directly electrospun into nanofibers for tissue engineering applications. Especially some synthetic polymers, such as PCL, poly(lactic acid) and poly(lactic-*co*-glycolic acid) (PLGA), could offer the scaffold sufficient mechanical property for hard tissue engineering, e.g., bone, which cannot be achieved by natural polymers. Highly aligned nanofibrous scaffolds, which favor the cell adhesion and proliferation, can be fabricated through post-drawing or using special collectors. However, most of these synthetic polymers are hydrophobic (5).

Compared to electrospinning, microfluidic spinning is more suitable for natural polymers than electrospinning. Microfluidic spinning has advantages such as a mild spinning environment, the ability to spin natural polymers without using synthetic polymers as aids and a toxic solvent, uniformity in the size and shape of the resulting fibers without the formation of droplets, and the ability to handle single fiber. Actually, most natural polymers that can be spun into fibers using microfluidic spinning are fabricated in an aqueous environment, hence avoiding denaturalization during the fabrication process (5).

Functional materials such as silver nanoparticles and growth factors have been co-electrospun with polymers to improve the antibacterial properties and healing efficiency of the mat. Electrospun mats hold great promise in skin tissue engineering due to their extracellular matrix-like structure (5). The migration of keratinocytes has been shown to be significantly stimulated on the collagen-gel-coating PCL mats (7).

Various extracellular matrix proteins, such as collagen, fibrin, or agarose, were encapsulated in alginate fibers to provide a suitable microenvironment for the cells and reconstruct fiber-shaped functional tissues to mimic the physiological environment (8).

4.1.2.2 *Treatment of Human Skin Wounds*

Human skin wounds involve a multitude of endogenous biochemical events and cellular reactions of the immune system (9). The healing process is extremely complex and affected by the physiological conditions of the patient, potential implications like infectious pathogens and inflammation as well as external factors. Due to increasing incidence of chronic wounds and proceeding resistance of infection pathogens, there is a strong need for effective therapeutic wound care.

Electrospun fibers with diameters in the nano- to micrometer range are gaining increasing interest. These resemble the structure of the native human extracellular matrix. Such fiber mats provide physical and mechanical protection, including protection against bacterial invasion.

Also, such fibers allow gas exchange and prevent the occlusion of the wound bed, thus facilitating wound healing. In addition, drugs can be incorporated within such fiber mats and their release can be adjusted by the material and dimensions of the individual fibers.

The current state of electrospun fibers for therapeutic application on skin wounds has been reviewed (9).

Antimicrobial peptides can be added to these systems. These aim to control the microbial proliferation and colonization of pathogens and to modulate the host's immune response (10).

Dressings that combine antimicrobial peptides with well-established polymeric dressings can be used in the treatment of acute and chronic wounds that are critically colonized or infected (11,12).

The antimicrobial performance of these dressings is dependent not only on the type of antimicrobial peptides immobilized but also on their activity and stability while functionalized. Thus, the importance of selecting the most appropriate immobilization process.

There are many strategies used to immobilize antimicrobial peptides onto electrospun dressings (13). The most common and simplest strategy is the co-spinning method. Here, antimicrobial peptides are immobilized as the polymeric nanofibers are produced, since one unique blend that incorporates all elements composing the dressing is electrospun at once.

The solubility of an antimicrobial peptide is a crucial physicochemical property on which its antimicrobial activity and target specificity depends. Antimicrobial peptides act on or enter through lipid membranes by solubilizing themselves in aqueous environments. When antimicrobial peptide molecules aggregate to prevent this action, their ability to interact with the cell membrane may be compromised.

The dosage of antimicrobial peptides to be delivered at site and its dissolution rate is controlled by selecting the most appropriate carrier with tunable activity. However, not all antimicrobial peptides should be used in this method (10).

4.1.2.3 *Regeneration of Functional Peripheral Nerve Tissue*

The regeneration of functional peripheral nerve tissue at critical-sized defect requires extracellular matrix analogues impregnated with appropriate biosignals to regulate the cell fate process and subsequent tissue progression (14). Electrospun aligned nanofibers have been developed as architectural analogues integrated with RADA16-I-BMHP1 as biofunctional peptides.

RADARADARADARADA (RADA 16-I) is a synthetic peptide of 16 amino acid residues, belonging to the family of self-complementary peptides. It consists of repeated segments of hydrophobic (alanine) and hydrophilic (arginine and aspartic acid) amino groups (15). This amphiphilic peptide was designed to self-assemble in a controlled way into fibrils and higher ordered structures depending on pH (16). BMHP1 is a bone marrow homing peptide (17).

Aligned PLGA-RADA16-I-BMHP1 nanofibers were fabricated and characterized for their *in vitro* potential using rat Schwann cell line and *in vivo* potential using a 10 *mm* sciatic nerve transection rat model (14).

PLGA-peptide scaffolds significantly promoted higher expression of genotypic markers and bipolar extension of Schwann cells. In addition, PLGA-peptide treated animals promoted the native collagen organization and remyelination and showed significantly higher recovery of sensorimotor and motor function than PLGA-treated groups ($p < 0.05$). The results of this study demonstrated that self-assembling peptide nanostructures on aligned PLGA nanofibers provided a better cell-matrix communication with significant functional restoration of the sciatic nerve (14).

4.1.3 *Functional Fluorophores*

Functional fluorophores are powerful tools for the study of biomolecules and cells in a noninvasive manner. The most recent advances in the design, preparation, and fine-tuning of fluorophores have been reviewed (18).

Examples of functional fluorophores that can be obtained by multicomponent reactions are summarized in Table 4.1. Some functional fluorophores are shown in Figure 4.1.

Cycloaddition reactions are a valuable strategy to access highly functionalized structures owing to their experimental ease, good synthetic yields, and compatibility with multiple functional groups (19) with the alkyne-azide 1,3-dipolar Huisgen cycloaddition (20) being one of the most widely used reactions in chemical biology.

This strategy was utilized to conjugate azidocoumarin, c.f. Figure 4.2, derivatives to an alkyne-containing isoindoline, c.f. Figure 4.3, nitroxyl in a copper-catalyzed azide-alkyne cycloaddition process to generate fluorophores with high sensitivity to oxidative processes (21).

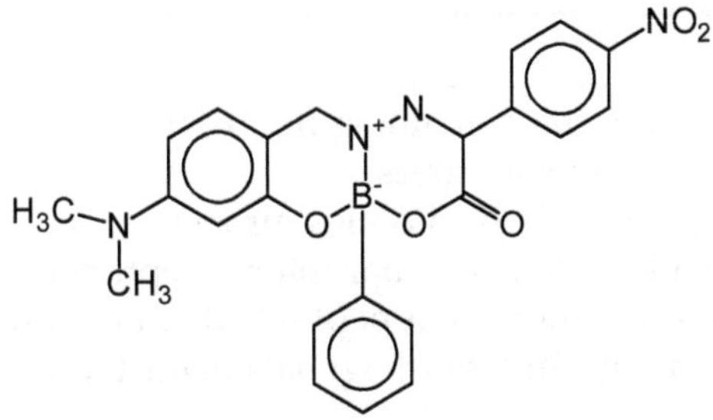

Polarity-dependent boron-containing fluorophore

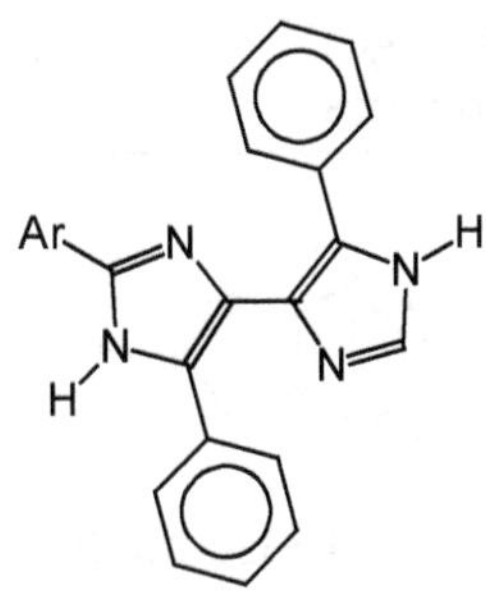

Bis(imidazole) heavy-metal probes

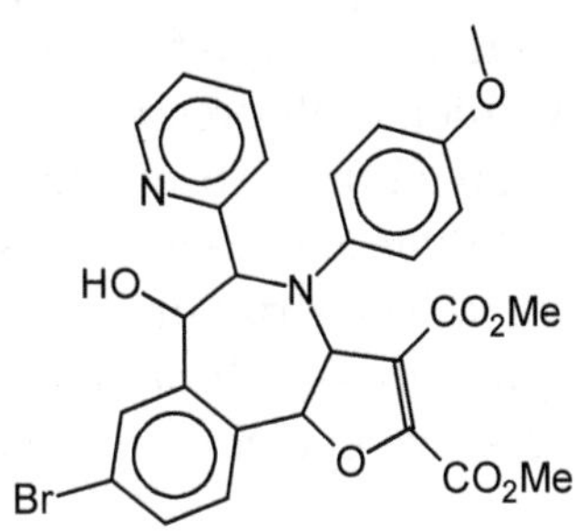

N-Heterocyclic carbene-derived optical probes

Figure 4.1 Functional fluorophores (18).

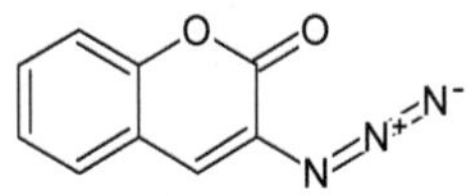

Figure 4.2 Azidocoumarin.

Table 4.1 Functional fluorophores (18).

Compound
Polarity-dependent boron-containing fluorophore
Bis(imidazole) heavy metal probes
N-Heterocyclic carbene-derived optical probes
Fluorophores that bind to benzodiazepine receptors in mitochondria
Coumarin-containing fluorescent peptoids that target mitochondria
Rhodamine-based tags for bioorthogonal chemistry and protein profiling
Dansyl-based protein reactive polymers
PhagoGreen as a pH-sensitive BODIPY fluorophore for *in vivo* imaging of phagocytic macrophages
Blue-emitting fluroisoquinolines

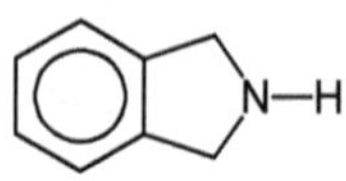

Figure 4.3 Isoindoline.

Another example is the preparation of highly decorated squaraine rotaxane dendrimers (22). Squaraine dyes are a class of organic dyes that show intense fluorescence, typically in the red and near infrared region (23). Absorption maxima are found between 630 *nm* and 670 *nm* and their emission maxima are between 650 *nm* and 700 *nm*. The structure of the squarylium dye III (2-[4-(dimethylamino)phenyl]-4-(4-dimethylazaniumylidenecyclohexa-2,5-dien-1-ylidene)-3-oxocyclobuten-1-olate) is shown in Figure 4.4.

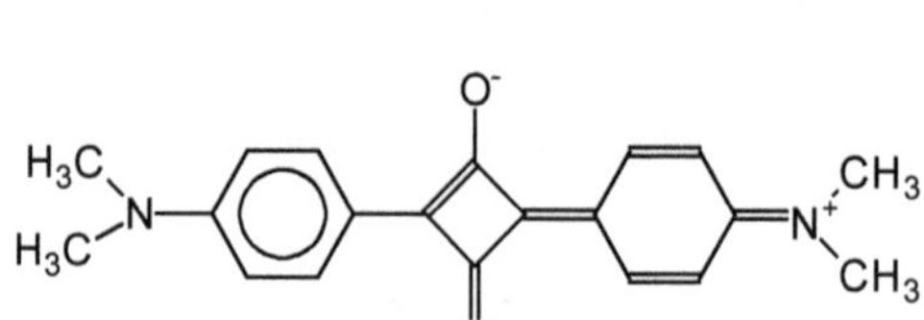

Figure 4.4 Squarylium dye III.

In a study (22), a squaraine fluorescent core was encapsulated within a macrocycle containing four alkyne groups that were clicked to azido amines to achieve bright deep-red fluorophores with a high photostability.

The photoclick reaction between tetrazoles and alkenes to generate environmentally sensitive fluorophores is shown in Figure 4.5.

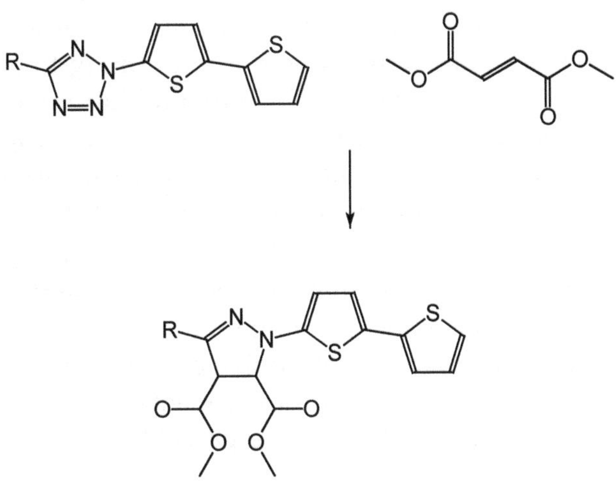

Figure 4.5 Photoclick reaction (18).

Functional fluorophores have been broadly applied to investigate different biomolecules, from small-molecule drugs to large enveloped viruses (24).

Metabolic signaling is an important area in the life sciences, which has become much more accessible due to the development of bioorthogonal functional fluorophores. After a study with metabolically compatible glycans (25), the metabolic incorporation of UDP-4-azido-4-de oxyxylose (UDP = uridine diphosphate) was investigated to study the function of glycosaminoglycans in zebrafish embryogenesis (26). The *in vivo* coupling of these sugars to fluorescent cyclooctynes revealed new links between glycosaminoglycan abundance and embryonic development (18).

Additional work has resulted in the adaptation of cycloaddition reactions for advanced imaging technologies, such as two-photon and fluorescence lifetime imaging (27), and the ratiometric visualization of alkyne-modified metabolites in living cells (28).

4.1.4 Protein Affinity Reagents

Protein affinity reagents are widely used in basic research, diagnostics and separations and for clinical applications, the most common of which are antibodies (29). However, often they suffer from high cost, and difficulties in their development, production and storage.

A synthetic polymer nanoparticle has been developed that can be engineered to have many of the functions of a protein affinity reagent. Polymer nanoparticles with a *nM* affinity to a key vascular endothelial growth factor (VEGF165) inhibit the binding of the signaling protein to its receptor VEGFR-2, preventing receptor phosphorylation and downstream VEGF165-dependent endothelial cell migration and invasion into the extracellular matrix.

In addition, the nanoparticles inhibit VEGF-mediated new blood vessel formation in matrigel plugs *in vivo*. The nontoxic nanoparticles were not found to exhibit off-target activity. These results support the assertion that synthetic polymers offer a new paradigm in the search for abiotic protein affinity reagents by providing many of the functions of their protein counterparts (29).

4.1.5 Lysozyme-Imprinted Polymers

Lysozyme-imprinted polymers have been prepared that bear modifiable sites within the imprinted cavity to introduce various functional groups via post-imprinting modifications (30). For this purpose, ([2-(2-methacrylamido)-ethyldithio]-ethylcarbamoyl-methoxy)acetic acid (MDTA), c.f. Figure 4.6, which has a carboxy group to interact with the target protein, lysozyme, and a disulfide linkage for post-imprinting modifications, was used as a functional monomer.

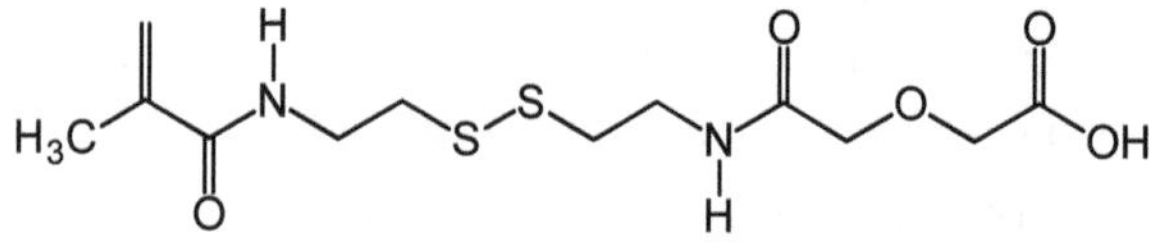

Figure 4.6 ([2-(2-Methacrylamido)-ethyldithio]-ethylcarbamoyl-methoxy)acetic acid.

A lysozyme-imprinted polymer film was prepared by the copolymerization of MDTA with a crosslinking agent, *N,N′*-methylenebisacrylamide, c.f. Figure 4.7, in the presence of lysozyme.

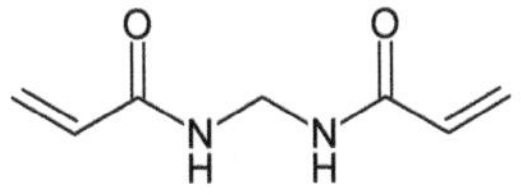

Figure 4.7 *N,N′*-Methylenebisacrylamide.

After removing lysozyme, followed by reducing the disulfide linkage, various functional groups, such as carboxy, amino, sulfonate, and oligo ethylene oxide, could be introduced to the exposed thiol groups via a disulfide exchange reaction using the pyridyldisulfide derivatives of these functional groups.

Various functional groups could be introduced reversibly via this post-imprinting disulfide exchange reaction after the construction of the lysozyme-imprinted cavities. The modification regulated the protein-binding activity. The proposed post-imprinting modification system, based on a molecular imprinting process, is expected to lead to the development of advanced materials for fine-tuning and/or introducing the desired functions (30).

4.1.6 *Glycoprotein-Functionalized Polymers*

Glycoproteins have interesting properties for medical uses such as biodegradability, biocompatibility, nontoxicity, antimicrobial and adsorption properties (31).

The materials are blended with different polymers such as chitosan, carboxymethyl cellulose, poly(*N*-vinyl-2-pyrrolidone), poly(caprolactone) (PCL), heparin, poly(styrene) (PS) fluorescent nanoparticles (PS-NPs) and carboxyl pullulan (PC), c.f. Figure 4.8, to improve their properties like thermal stability, mechanical properties, resistance to pH, chemical stability and toughness.

A review has been presented concerning the synthesis and characterization of blends and composites of glycoproteins with natural and synthetic polymers (31). Also, their potential applications in the biomedical field have been discussed, such as drug delivery system, insulin delivery, antimicrobial wound dressing uses, targeting

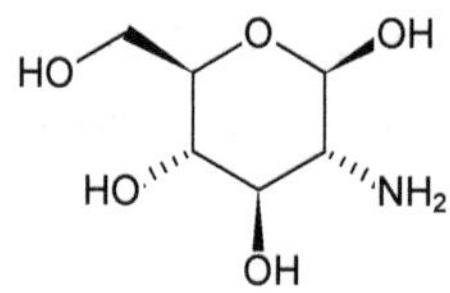

Chitosan

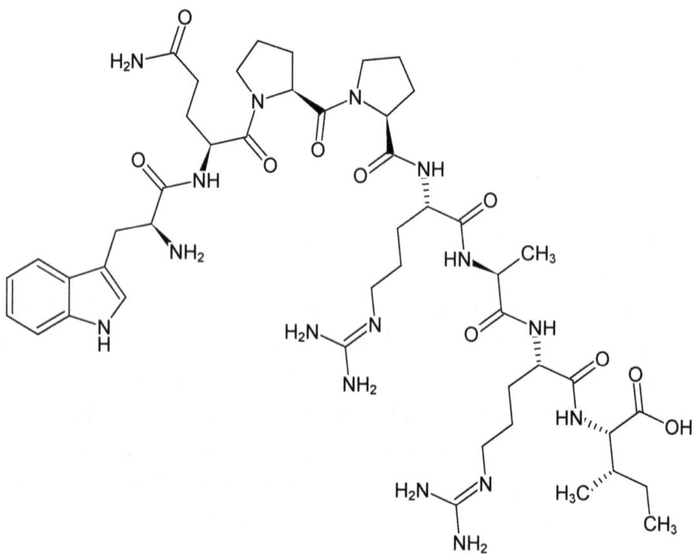

Heparin

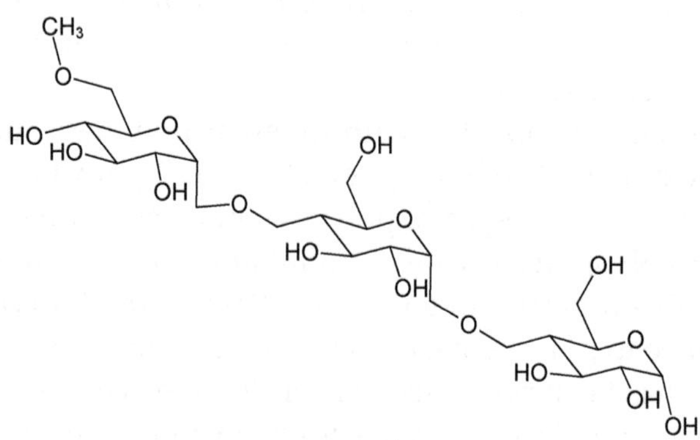

Pullulan

Figure 4.8 Materials for blending.

of cancer cells, development of anticancer vaccines, development of new biopolymers, glycoproteome research, food product and detection of dengue glycoproteins (31).

4.1.6.1 Biomolecular Self-Assembly

Artificial glycoclusters based on synthetic polymers which show a multivalent glycocluster effect have been developed, not only as a model of molecular recognition involving glycolipids and glycoproteins, but also as a cell culture substrate, material for trapping viruses and toxins, sensing material, drug delivery material, and regulating material for amyloid fibrils (32).

In densely arranged glycoclusters based on synthetic homopolymers, such as PS and poly(acrylamide) as the main chain, the lectin-recognizing ability may decrease due to steric hindrance in some cases (33).

Although copolymerization can space the oligosaccharides at average intervals, it is difficult to synthesize macromolecules in which the intervals of the oligosaccharides are regularly controlled (34).

The deoxyribonucleic acid (DNA) double helix has a relatively rigid conformation and regular helical pitch. Therefore, it is promising as a scaffold for arranging functional molecules at regular intervals.

Galactose-modified oligodeoxyribonucleotides were synthesized by solid-phase phosphoramidite chemistry using a 5′-β-galactoside modified deoxyuracil derivative, which was synthesized via Sonogashira coupling (35).

The Sonogashira reaction is a cross-coupling reaction used in organic synthesis to form carbon-carbon bonds (36,37). It uses a palladium catalyst to form a carbon-carbon bond between a terminal alkyne and an aryl or vinyl halide.

In eukaryotes, DNA is packaged in a nucleus as chromatin by the interaction with histone proteins, in which gene expression is usually repressed. When chromatin remodeling occurs, such as acetylation of histones, the gene expression is turned on. Artificial on-off switching of gene expression is important in synthetic biology, especially for medical applications.

An artificial photoresponsive regulation system of transcription was developed using an azobenzene-modified promoter (38). Also, a phosphorylation signal-responsive artificial gene regulation system has been developed using polymers carrying oligopeptides (39).

Another strategy for an artificial on-off switching system of gene expression is its use for the recognition of plasmid-lactose conjugate to lectin (40,41).

Conjugates of plasmid DNAs with lactose were prepared by diazo-coupling with a diazonium derivative of lactose (42,43). The plasmid-lactose conjugate retained the transcription activity, despite the existence of nucleotides modified with lactose.

When the plasmid-lactose conjugate was complexed strongly with RCA_{120} lectin, access of RNA polymerase to the template DNA was hindered and the transcription of DNA was repressed. But when an excess amount of lactose or lactose-carrying polymer was added to the complex, the binding between the plasmid-lactose conjugate and lectin was relaxed, and RNA polymerase became accessible to the DNA to recover the transcription (34).

The construction of virus-like nanostructures from peptides designed by mimicking the self-assembly strategy of natural viral capsids has been studied (44–47).

It has been reported that even short peptides, such as dipeptides and tripeptides, function as self-assembly units. Oxidized glutathione formed fibrous assemblies in organic solvents yielded an organogel (48). Glutathione is shown in Figure 4.9.

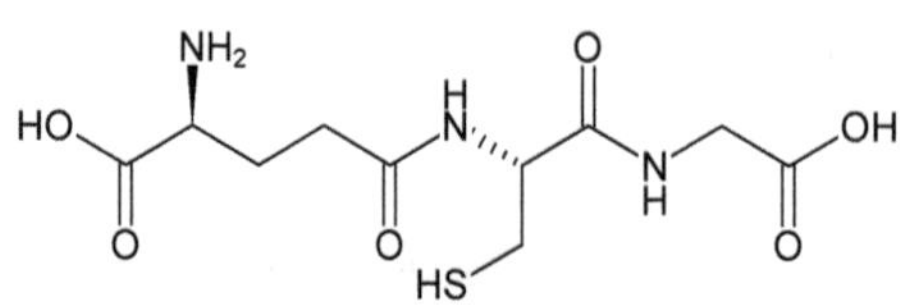

Figure 4.9 Glutathione.

Later, it was found that trigonal conjugation of glutathione forms spherical assemblies with sizes of 100–250 *nm* in water (49).

Scanning electron microscopy studies of 1 *mM* trigonal glutathione showed the formation of wrinkly collapsed spherical assemblies, thus suggesting the existence of a hollow interior. Some guest molecules such as uranine and methyl orange were encapsulated into the trigonal glutathione nanosphere. When dithiothreitol was added to the guest-encapsulated trigonal glutathione nanospheres as a reducing agent, the recombination of disulfide bonds occurred and glutathione was released while maintaining the nanosphere structure. As a result, guest molecules were gradually released. The above-mentioned compounds are shown in Figure 4.10

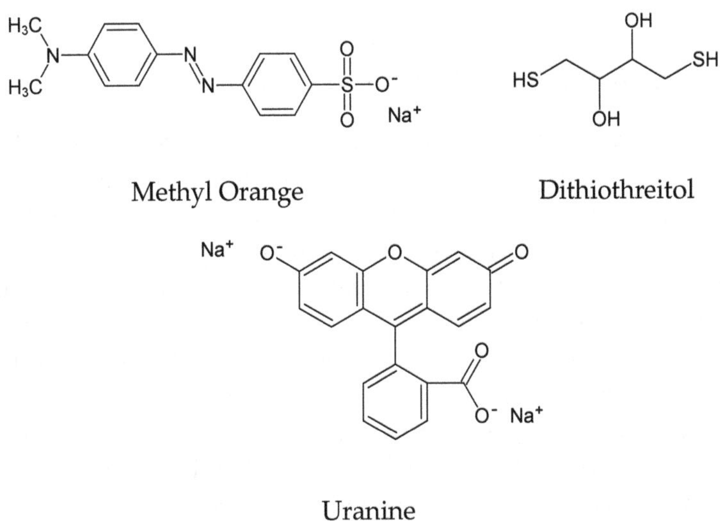

Figure 4.10 Compounds for trigonal glutathione nanospheres.

In order to construct more regular nano-assemblies from glutathione derivatives, 1,3,5-tris(aminomethyl)-2,4,6-triethylbenzene, c.f. Figure 4.11, was employed as the core of a new trigonal glutathione (50). The glutathione arms on the 1,3,5-positions were assembled on one side of the scaffold due to the alternative conformation of the core. The conformation-regulated trigonal glutathione self-assembled into hard spherical structures with sizes of 310 ± 50 *nm*, which had regular morphologies and enhanced rigidity compared with those formed from conformationally non-regulated trigonal glutathione.

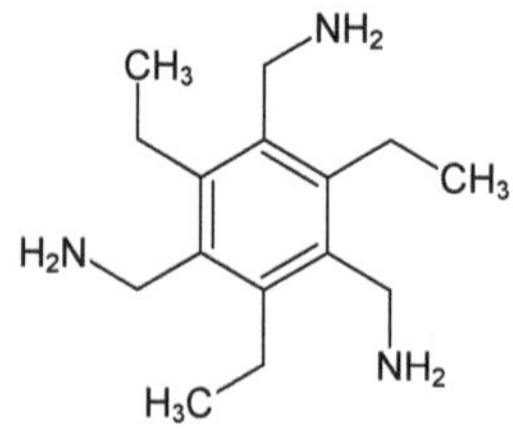

Figure 4.11 1,3,5-Tris(aminomethyl)-2,4,6-triethylbenzene.

4.1.7 Biopolymer-Based Functional Composites

Many biopolymer-based functional composites have been developed to increase the value of raw biopolymers that are obtained from natural sources or microbial systems. Uses for various biopolymers have been reviewed, along with important methods for biopolymer-based composites preparation, and surface topography for tissue engineering (51).

Also, functional biopolymer composites have been detailed that are used in various medical applications, such as tissue engineering comprising skin, bone, cartilage, vascular graft, and other organs, implantable medical devices including stent and barrier membrane, and some delivery systems of bioactive agents.

4.1.7.1 Collagen

Collagen is the major structural component in connective tissues such as tendon, skin and bone. It is widely used as a natural biopolymer in different formats, including sponges, films, membranes, and tubes (52,53).

Collagen is a typical polyampholyte and has an isoelectric point near the physiological pH, which means that it does not swell or dissolve in neutral pH solution. In order to improve the water-solubility of collagen, an attempt has been made to introduce additional carboxylic groups (54).

The superior water-solubility of polyanionic collagen ensures that a clear solution can be obtained at physiological pH, which makes it an ideal candidate for injectable biomaterials.

It has been demonstrated that a succinoylated collagen has a denaturation temperature of 34.7°C, which is lower than that of the native collagen, i.e., 38.4°C.

In an improved method for collagen functionalization, a poly(γ-glutamic acid) derivative was used, i.e., poly(γ-glutamic acid)/carbodiimide/*N*-hydroxysuccinimide (54). This compound is shown in Figure 4.12.

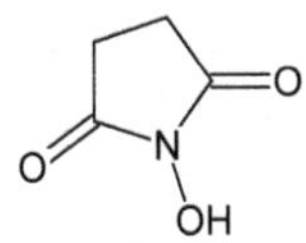

Figure 4.12 *N*-Hydroxysuccinimide.

The modified collagen could be dissolved in the neutral pH water to form a clear solution due to the decrease of the isoelectric point. The modified collagen had a significantly improved denaturation temperature, which was at least 6.0°C higher than that of the native collagen (54).

In another study, a functional silane, 3-glycidoxypropyl trimethoxysilane, c.f. Figure 4.13, was mixed with a diluted collagen solution under acidic conditions, then cast and dried (55).

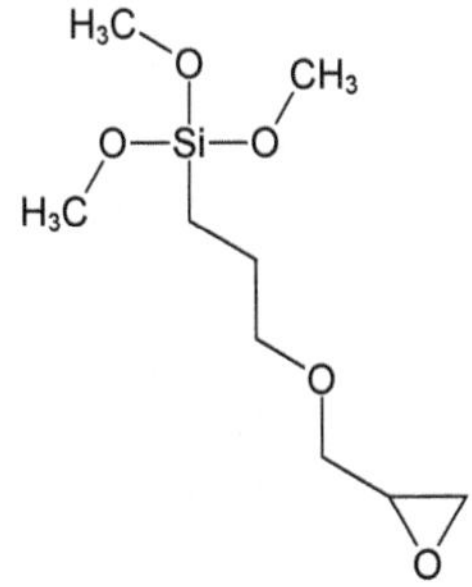

Figure 4.13 3-Glycidoxypropyl trimethoxysilane.

This strategy is similar to a previous study (56), where the epoxy group of 3-glycidoxypropyl trimethoxysilane should form a covalent bond with the protein backbone, while the silanol groups can be involved in the hydrogel crosslinking (57).

4.1.7.2 Calcium Phosphate Composites

Calcium phosphate-containing materials have many peculiar and intriguing properties. In nature, calcium phosphate is found in nanostructured form embedded in a soft proteic matrix as the main mineral component of bones and teeth.

Some of the guidelines for the synthesis of calcium phosphate composites have been reviewed (58). The extraordinary stoichiometric flexibility, the different stabilities exhibited by its different forms as a function of pH and the highly dynamic nature of its surface ions render calcium phosphate one of the most versatile materials for nanomedicine.

Calcium phosphate has been used together with biologically active polymers or prepared within a thermo-responsive and magneto-responsive hydrogel, respectively.

Beside small organic acids, polymers bearing carboxylic functionalities were also used to template the synthesis of calcium phosphates. Some examples are poly(acrylic acid) (59,60), poly(aspartic acid) (61), or block copolymers with a poly(acrylic) block (62).

When polymeric molecules are used as templating agents, the crystallinity of the calcium phosphate nuclei decreases and often the product of the reaction in the nucleation stage is kinetically favored amorphous calcium phosphate (58).

The surface functionalization with biologically active molecules, such as peptides, proteins, or targeting groups, may occur via non-covalent interactions such as electrostatic forces or van der Waals interactions.

In the field of tissue engineering, synthetic calcium phosphates are of special interest for bone regeneration because of their chemical and crystallographic similarity with natural inorganic components of bones.

The chemical modification of hydrogel polymeric bone through the introduction of functional groups that are able to bind Ca^{2+} is a strategy that has been explored over the last decade (58).

An environmentally friendly and bioactive material has been prepared from a cellulose grafted soy protein isolate for biomimetic calcium phosphate mineralization (63). Neat cellulose was oxidized using periodate, then chemically modified by reacting with soy protein isolate followed by soaking in a doubly concentrated simulated body fluid solution.

The cytotoxicity of the cellulose/SPI/calcium phosphate hybrid was evaluated using animal fibroblast baby hamster kidney cells. The cytotoxicity results suggested cellulose/soy protein isolate/calcium phosphate hybrid as potentially useful scaffold for regenerative therapies (63).

4.1.7.3 *Poly(γ-glutamic acid)*

Poly(γ-glutamic acid) is an unusual anionic polypeptide containing D- and/or L-glutamic acid units polymerized through amide linkages between α-amino and γ-carboxylic acid groups. L-glutamic acid is shown in Figure 4.14.

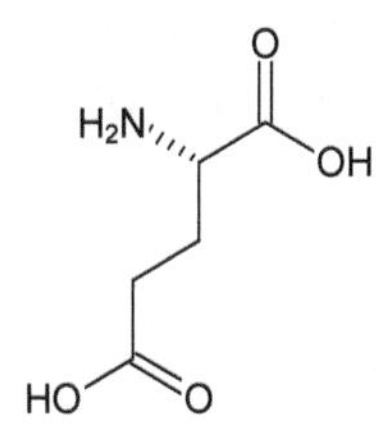

Figure 4.14 L-Glutamic acid.

Poly(γ-glutamic acid) can be degraded by enzyme actions. Three stereochemically different types of poly(γ-glutamic acid) have been reported: A homopolymer composed of *D*-glutamate, a homopolymer composed of L-glutamate, and a copolymer containing D- and L-glutamate units arranged randomly (64)

Poly(γ-glutamic acid) contains repeating units of a pendant carboxylic acid group that can be used for functionalization (65,66)

4.1.7.4 *Hyaluronic Acid*

Hyaluronic acid is an essential functional component of almost all tissues in vertebrates. Therefore, multifarious animal tissues, e.g., rooster combs, shark skin, and bovine eye balls, have been used for isolating and producing hyaluronic acid with a high molar mass (67).

4.1.7.5 *Insulin-Secreting Clusters*

Transplantation of islets isolated *in vitro* from a donor pancreas has the potential to become a widely applicable treatment for insulin-dependent diabetes (68).

Functionalized 3D silk matrices were shown for the generation of insulin-secreting islet-like clusters from mouse and human primary cells (69). Conventional molecular cloning techniques were used to insert short peptides at the N-terminus (RGD, RGE, FN and VN) or in the repetitive part (2R and 3R) of the recombinant spider silk protein 4RepCT. Significantly more clusters were formed on silk foam with the cell binding motif Arg-Gly-Asp (RGD) compared to wild type silk. The obtained clusters formed on the silk matrix maintain functional insulin release upon

of transplanted clusters showed engraftment with increasing vessel formation during time. The size of the clusters increased over time without cell toxicity (69).

4.2 pH-Sensitive Polymers

The pH-sensitive polymers are polyelectrolytes that contain weak acidic or basic groups that either accept or release protons in response to changes in the environmental pH (70).

The pendant acidic or basic groups on the polyelectrolytes are undergoing an ionization of their acidic or basic groups of monoacids or monobases. However, the complete ionization on polyelectrolytes is more difficult due to electrostatic effects exerted by other adjacent ionized groups. This makes the apparent dissociation constant (K_a) different from that of the corresponding monoacid or the corresponding monobase.

The pH range in which a reversible phase transition occurs can be controlled using two strategies (70):

1. Selecting the ionizable moiety with a *pKa* matching the desired pH range. Therefore, the proper selection between

polyacid or polybase should be considered for the desired application.

2. Incorporating hydrophobic moieties into the polymer backbone and controlling their nature, amount and distribution. When ionizable groups become neutral, i.e., nonionized, and the electrostatic repulsion forces disappear within the polymer network, hydrophobic interactions are dominating. The introduction of a more hydrophobic moiety may offer a more compact conformation in the uncharged state and a more charged phase transition. The hydrophobicity of these polymers can be controlled by the copolymerization of hydrophilic ionizable monomers with more hydrophobic monomers with or without pH-sensitive moieties such as 2-hydroxyethyl methacrylate, methyl methacrylate and maleic anhydride.

Polyacidic polymers will be unswollen at low pH, since the acidic groups will be protonated and unionized. When the pH increases, a negatively charged polymer will start to swell. The opposite behavior is found in polybasic polymers, since the ionization of the basic groups will increase when the pH decreases.

Examples of pH-sensitive polymers with anionic groups are poly(carboxylic acid)s, such as poly(acrylic acid) or poly(methacrylic acid). Another type of polyacidic polymer is poly(sulfonamide)s. An example is shown in Figure 4.15.

Examples of cationic polyelectrolytes are poly(*N*,*N*-dialkyl aminoethyl methacrylate)s, poly(lysine), poly(ether imide), and chitosan.

Some pH-sensitive polymers can be crosslinked to form hydrogels (70). Their behavior is not only influenced by the nature of the ionizable groups, the polymer composition, and the hydrophobicity of the polymeric backbone, but also by the crosslinking density. The higher the crosslinking density, the lower the permeability. This is especially significant in the case of high molecular weight solutes. Most polyelectrolyte gels show a decrease in their modulus with an increasing degree of swelling.

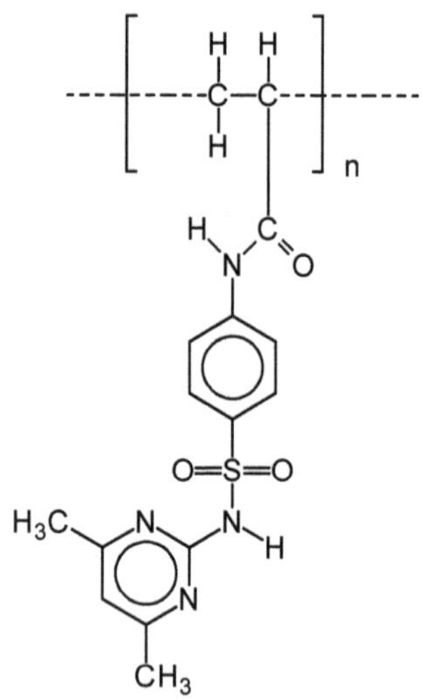

Figure 4.15 4-Amino-*N*-[4,6-dimethyl-2-pyrimidinyl]-benzenesulfonamine (sulfomethazine) containing polymer (71).

4.2.1 *Drug and Gene Delivery Systems*

Several biomedical applications have used pH-sensitive polymers, most importantly in drug and gene delivery systems and glucose sensors.

In the human body, the pH varies along the gastrointestinal tract between pH 2 in the stomach and pH 10 in the colon. This condition makes pH-sensitive polymers ideal for a colon-specific drug delivery. The most common approach utilizes enteric polymers that resist degradation in acidic environments.

4.2.2 *Insulin Delivery Systems*

One of the most popular applications of pH-sensitive polymers is the fabrication of insulin delivery systems for the treatment of diabetic patients (70). The delivery of insulin is different from delivering other drugs, since insulin has to be delivered in an exact amount at the exact time needed. Many devices have been developed for this purpose and all of them have a glucose sensor built into the system. In a glucose-rich environment, such as the bloodstream after a meal, the oxidation of glucose to gluconic acid catalyzed by glucose oxidase can lower the pH to approximately 5.8. This enzyme is probably the most widely used in glucose sensing, making it possible to apply different types of pH-sensitive hydrogels for modulated insulin delivery (72).

References

1. F. Seidi, R. Jenjob, and D. Crespy, *Chemical Reviews*, Vol. 118, p. 3965, 2018.
2. H. Tian, Z. Tang, X. Zhuang, X. Chen, and X. Jing, *Progress in Polymer Science*, Vol. 37, p. 237, 2012. 37/2 Topical Issue on Biomaterials.
3. X. Lou, C. Detrembleur, and R. Jérôme, *Macromolecular Rapid Communications*, Vol. 24, p. 161, 2003.
4. J. Niu, D.J. Lunn, A. Pusuluri, J.I. Yoo, M.A. O'Malley, S. Mitragotri, H.T. Soh, and C.J. Hawker, *Nature Chemistry*, Vol. 9, p. 537, January 2017.
5. J. Cheng, Y. Jun, J. Qin, and S.-H. Lee, *Biomaterials*, Vol. 114, p. 121, 2017.
6. L.W. Chow, *Electrospinning functionalized polymers for use as tissue engineering scaffolds* in K. Chawla, ed., *Biomaterials for Tissue Engineering: Methods and Protocols*, Vol. 1758 of *Methods in Molecular Biology*, pp. 27–39. Springer New York, New York, NY, 2018.
7. X. Fu, M. Xu, J. Liu, Y. Qi, S. Li, and H. Wang, *Biomaterials*, Vol. 35, p. 1496, 2014.
8. H. Onoe, T. Okitsu, A. Itou, M. Kato-Negishi, R. Gojo, D. Kiriya, K. Sato, S. Miura, S. Iwanaga, K. Kuribayashi-Shigetomi, Y.T. Matsunaga, Y. Shimoyama, and S. Takeuchi, *Nature Materials*, Vol. 12, p. 584, March 2013.
9. J. Wang and M. Windbergs, *European Journal of Pharmaceutics and Biopharmaceutics*, Vol. 119, p. 283, 2017.
10. H.P. Felgueiras and M.T.P. Amorim, *Colloids and Surfaces B: Biointerfaces*, Vol. 156, p. 133, 2017.
11. F. Gottrup, J. Apelqvist, T. Bjarnsholt, R. Cooper, Z. Moore, E.J. Peters, and S. Probst, *Journal of Wound Care*, Vol. 23, p. 477, 2014.
12. F.D. Halstead, M. Rauf, A. Bamford, C.M. Wearn, J.R.B. Bishop, R. Burt, A.P. Fraise, N.S. Moiemen, B.A. Oppenheim, and M.A. Webber, *Burns*, Vol. 41, p. 1683, 2015.
13. J.-B.D. Green, T. Fulghum, and M.A. Nordhaus, *Chem. Rev*, Vol. 109, p. 5437, 2009.
14. M. Nune, A. Subramanian, U.M. Krishnan, S.S. Kaimal, and S. Sethuraman, *Nanomedicine*, Vol. 12, p. 219, 2017.

15. H. Yokoi, T. Kinoshita, and S. Zhang, *Proceedings of the National Academy of Sciences*, Vol. 102, p. 8414, 2005.
16. P. Arosio, M. Owczarz, H. Wu, A. Butté, and M. Morbidelli, *Biophysical Journal*, Vol. 102, p. 1617, 2012.
17. F.-Y. Cao, W.-N. Yin, J.-X. Fan, R.-X. Zhuo, and X.-Z. Zhang, *Biomater. Sci.*, Vol. 3, p. 345, 2015.
18. F. de Moliner, N. Kielland, R. Lavilla, and M. Vendrell, *Angewandte Chemie International Edition*, Vol. 56, p. 3758, 2017.
19. N. Nishiwaki, ed., *Methods and Applications of Cycloaddition Reactions in Organic Syntheses*, John Wiley & Sons, Hoboken, 2013.
20. F. Himo, T. Lovell, R. Hilgraf, V.V. Rostovtsev, L. Noodleman, K.B. Sharpless, and V.V. Fokin, *Journal of the American Chemical Society*, Vol. 127, p. 210, 2005.
21. J.C. Morris, J.C. McMurtrie, S.E. Bottle, and K.E. Fairfull-Smith, *The Journal of Organic Chemistry*, Vol. 76, p. 4964, 2011.
22. S. Xiao, N. Fu, K. Peckham, and B.D. Smith, *Organic Letters*, Vol. 12, p. 140, 2010.
23. Wikipedia contributors, Squaraine dye — Wikipedia, the free encyclopedia, https://en.wikipedia.org/w/index.php?title=Squaraine_dye&oldid=813219928, 2017. [Online; accessed 30-August-2018].
24. L.-L. Huang, G.-H. Lu, J. Hao, H. Wang, D.-L. Yin, and H.-Y. Xie, *Analytical Chemistry*, Vol. 85, p. 5263, 2013.
25. N.J. Agard, J.A. Prescher, and C.R. Bertozzi, *Journal of the American Chemical Society*, Vol. 126, p. 15046, 2004.
26. P. Agarwal, B.J. Beahm, P. Shieh, and C.R. Bertozzi, *Angewandte Chemie*, Vol. 127, p. 11666, 2015.
27. B. Belardi, A. de la Zerda, D.R. Spiciarich, S.L. Maund, D.M. Peehl, and C.R. Bertozzi, *Angewandte Chemie International Edition*, Vol. 52, p. 14045, 2013.
28. H. Fu, Y. Li, L. Sun, P. He, and X. Duan, *Anal. Chem.*, Vol. 87, p. 11332, 2015.
29. H. Koide, K. Yoshimatsu, Y. Hoshino, S.-H. Lee, A. Okajima, S. Ariizumi, Y. Narita, Y. Yonamine, A.C. Weisman, Y. Nishimura, N. Oku, Y. Miura, and K.J. Shea, *Nature Chemistry*, Vol. 9, p. 715, March 2017.
30. H. Sunayama, Y. Kitayama, and T. Takeuchi, *Journal of Molecular Recognition*, Vol. 31, p. e2633, 2017.
31. S. Tabasum, A. Noreen, A. Kanwal, M. Zuber, M.N. Anjum, and K.M. Zia, *International Journal of Biological Macromolecules*, Vol. 98, p. 748, 2017.
32. Y. Miura, Y. Hoshino, and H. Seto, *Chemical Reviews*, Vol. 116, p. 1673, 2016.
33. T. Hasegawa, S. Kondoh, K. Matsuura, and K. Kobayashi, *Macromolecules*, Vol. 32, p. 6595, 1999.

34. K. Matsuura, *Bulletin of the Chemical Society of Japan*, Vol. 90, p. 873.884, 2017.
35. M. Tsukamoto and Y. Hayakawa, *Frontiers in Organic Chemistry*, Vol. 1, p. 3, 2005.
36. K. Sonogashira, Y. Tohda, and N. Hagihara, *Tetrahedron Letters*, Vol. 16, p. 4467, 1975.
37. Wikipedia contributors, Sonogashira coupling — Wikipedia, the free encyclopedia, https://en.wikipedia.org/w/index.php?title=Sonogashira_coupling&oldid=855490506, 2018. [Online; accessed 31-August-2018].
38. H. Asanuma, D. Tamaru, A. Yamazawa, M. Liu, and M. Komiyama, *ChemBioChem*, Vol. 3, p. 786, 2002.
39. Y. Katayama, K. Fujii, E. Ito, S. Sakakihara, T. Sonoda, M. Murata, and M. Maeda, *Biomacromolecules*, Vol. 3, p. 905, 2002.
40. K. Matsuura, K. Hayashi, and K. Kobayashi, *Chem. Commun.*, pp. 1140–1141, 2002.
41. K. Matsuura, K. Hayashi, and K. Kobayashi, *Biomacromolecules*, Vol. 6, p. 2533, 2005.
42. T. Akasaka, K. Matsuura, N. Emi, and K. Kobayashi, *Biochemical and Biophysical Research Communications*, Vol. 260, p. 323, 1999.
43. K. Matsuura, T. Akasaka, M. Hibino, and K. Kobayashi, *Bioconjugate Chemistry*, Vol. 11, p. 202, 2000.
44. K. Matsuura, K. Murasato, and N. Kimizuka, *Journal of the American Chemical Society*, Vol. 127, p. 10148, 2005.
45. K. Murasato, K. Matsuura, and N. Kimizuka, *Biomacromolecules*, Vol. 9, p. 913, 2008.
46. K. Matsuura, H. Hayashi, K. Murasato, and N. Kimizuka, *Chem. Commun.*, Vol. 47, p. 265, 2011.
47. K. Matsuura, K. Murasato, and N. Kimizuka, *International Journal of Molecular Sciences*, Vol. 12, p. 5187, August 2011.
48. R.P. Lyon and W.M. Atkins, *Journal of the American Chemical Society*, Vol. 123, p. 4408, 2001.
49. K. Matsuura, H. Matsuyama, T. Fukuda, T. Teramoto, K. Watanabe, K. Murasato, and N. Kimizuka, *Soft Matter*, Vol. 5, p. 2463, 2009.
50. K. Matsuura, K. Fujino, T. Teramoto, K. Murasato, and N. Kimizuka, *Bulletin of the Chemical Society of Japan*, Vol. 83, p. 880, 2010.
51. S.-B. Park, E. Lih, K.-S. Park, Y.K. Joung, and D.K. Han, *Progress in Polymer Science*, Vol. 68, p. 77, 2017.
52. M.G. Patino, M.E. Neiders, S. Andreana, B. Noble, and R.E. Cohen, *Implant Dentistry*, Vol. 11, p. 280, 2002.
53. X. Yu, C. Tang, S. Xiong, Q. Yuan, Z. Gu, Z. Li, and Y. Hu, *Current Organic Chemistry*, Vol. 20, p. 1797, 2016.
54. M. Zhang, J. Yang, C. Ding, L. Huang, and L. Chen, *Reactive and Functional Polymers*, Vol. 122, p. 131 , 2018.

55. S. Chen, S. Chinnathambi, X. Shi, A. Osaka, Y. Zhu, and N. Hanagata, *J. Mater. Chem.*, Vol. 22, p. 21885, 2012.
56. A. Fatimi, J.F. Tassin, S. Quillard, M.A.V. Axelos, and P. Weiss, *Biomaterials*, Vol. 29, p. 533, 2008.
57. S. Heinemann, T. Coradin, and M.F. Desimone, *Biomaterials Science*, Vol. 1, p. 688, 2013.
58. F. Ridi, I. Meazzini, B. Castroflorio, M. Bonini, D. Berti, and P. Baglioni, *Advances in Colloid and Interface Science*, Vol. 244, p. 281, 2017. Special Issue in Honor of the 90th Birthday of Prof. Eli Ruckenstein.
59. E. Bertoni, A. Bigi, G. Falini, S. Panzavolta, and N. Roveri, *J. Mater. Chem.*, Vol. 9, p. 779, 1999.
60. A. Bigi, B. Bracci, G. Cojazzi, S. Panzavolta, and K. Rubini, *Journal of Biomaterials Science, Polymer Edition*, Vol. 15, p. 243, 2004.
61. S.-D. Jiang, Q.-Z. Yao, G.-T. Zhou, and S.-Q. Fu, *The Journal of Physical Chemistry C*, Vol. 116, p. 4484, 2012.
62. B.P. Bastakoti, M. Inuoe, S.-i. Yusa, S.-H. Liao, K.C.-W. Wu, K. Nakashima, and Y. Yamauchi, *Chem. Commun.*, Vol. 48, p. 6532, 2012.
63. A. Salama, N. Shukry, A. El-Gendy, and M. El-Sakhawy, *Industrial Crops and Products*, Vol. 95, p. 170, 2017.
64. E.C. King, A.J. Blacker, and T.D.H. Bugg, *Biomacromolecules*, Vol. 1, p. 75, 2000.
65. M.-H. Sung, C. Park, C.-J. Kim, H. Poo, K. Soda, and M. Ashiuchi, *The Chemical Record*, Vol. 5, p. 352, 2005.
66. A. Ogunleye, A. Bhat, V.U. Irorere, D. Hill, C. Williams, and I. Radecka, *Microbiology*, Vol. 161, p. 1, 2015.
67. G. Kogan, L. Šoltés, R. Stern, and P. Gemeiner, *Biotechnology Letters*, Vol. 29, p. 17, 2007.
68. A.M.J. Shapiro, J.R.T. Lakey, E.A. Ryan, G.S. Korbutt, E. Toth, G.L. Warnock, N.M. Kneteman, and R.V. Rajotte, *New England Journal of Medicine*, Vol. 343, p. 230, 2000.
69. N.D. Shalaly, M. Ria, U. Johansson, K. Åvall, P.-O. Berggren, and M. Hedhammar, *Biomaterials*, Vol. 90, p. 50, 2016.
70. M.R. Aguilar, C. Elvira, A. Gallardo, B. Vázquez, and J. Román, *Smart polymers and their applications as biomaterials* in N. Ashammakhi, R. Reis, and E. Chiellini, eds., *Topics in Tissue Engineering*, Vol. 3, chapter 6, pp. 1–27. Woodhead Publishing, an imprint of Elsevier Science, Sawston, UK, 2007.
71. B. Jeong, Y.H. Bae, and S.W. Kim, *Journal of Biomedical Materials Research*, Vol. 50, p. 171, 2000.
72. K. Podual, F.J. Doyle, and N.A. Peppas, *Polymer*, Vol. 41, p. 3975, 2000.

5

Pharmaceutical Applications

The administration of drugs is a main challenge for pharmaceutical and medicinal applications. It has certainly benefited from the application of synthetic polymers (1). However, despite an enormous effort to develop new materials for drug delivery applications, only a few of them have entered the market due to the hurdles of regulation, production, and cost efficiency.

The classes of synthetic polymers which are on the market, as well as the latest developments in clinical trials, have been summarized (1). Also, their application in polymer-drug conjugates as excipients in nano-/microscopic and macroscopic drug carriers as polymeric coatings or as polymeric drugs have been detailed.

Polymer classes that can be used for pharmaceutical applications are collected in Table 5.1.

5.1 Poly(ethylene glycol)

Poly(ethylene glycol) (PEG) is probably the most well-known poly-(ether) in pharmaceutical applications. A variety of functionalized PEGs are offered from commercial sources, with different α and/or ω functionalities, e.g., methoxy groups, amines, or thiols. Furthermore, side group functionalized PEG derivatives are known in the literature, e.g., functionalized with furfuryl groups (2–4).

An outstanding property of PEG is the so-called *stealth effect,* which was discovered in 1990 for modified liposomes (5,6).

247

Table 5.1 Polymer classes used for pharmaceutical applications (1).

Polymer class	Polymer class
Poly(ether)	Poly(ester)
Poly(carbonate)	Poly(amino acid)
Poly(*N*-acrylamide)	Poly(phosphoester)
Poly(siloxane)	Poly(vinyl alcohol)
Poly(vinyl ether)	Poly(*N*-vinyl amide)
Poly(acrylate)	Poly(cyanoacrylate)
Poly(acrylonitrile)	Poly(urethane)
Poly(styrene)	Poly(anhydride)

This is the ability to protect molecules or nanoscopic objects against unspecific interactions with blood components, e.g., opsonization, by attaching specific proteins to the surface (7).

5.1.1 *Carboxylic Acid Functionalization*

After activation by electrophilic groups, PEG derivatives are useful for coupling to nucleophilic groups, such as amino groups, of biologically active molecules. In particular, active esters and other carboxylic acid derivatives of PEG have been used to attach PEG to proteins bearing amino groups.

A persistent problem associated with the preparation of carboxyl-functionalized polymers has been the difficulty in obtaining the desired polymer product at a sufficiently high purity level (8).

For example, a method has been presented (9) of synthesizing PEG carboxylic acids, which consists of converting PEG with hydroxy end groups into an ethyl ester by the base-catalyzed reaction of PEG with hydroxy end groups with an α-halo ethyl ester, followed by the base-promoted hydrolysis of the ester. However, this approach provides PEG containing acids of only about 85% purity, with the main contaminant of PEG with hydroxy end groups, which cannot be separated from the PEG carboxylic acid using typical purification methods such as precipitation, crystallization or extraction. The removal of the impurity requires the use of preparative ion exchange column chromatography, which is time-consuming and expensive (8).

The preparation of PEG carboxylic acids has been reported (10,11) that contain little or no starting material, i.e., PEG alcohol, by employing a tertiary alkyl haloacetate to prepare a tertiary alkyl ester-functionalized PEG, which is then hydrolyzed with acid, preferably trifluoroacetic acid.

However, use of trifluoroacetic acid can result in purification and product stability problems. Trifluoroacetic acid is difficult to completely remove from the final carboxyl-functionalized polymer. The presence of residual trifluoroacetic acid results in poor product stability, due to degradation of the polymer caused by acid-promoted autoxidation. Also, it has been reported that acids catalyze the formation of hydroperoxides and hydroperoxide rupture, leading to the cleavage of polyoxyethylene chains (12).

A method for preparing a water soluble, non-peptidic polymer functionalized with a carboxyl group has been presented (8). The method consists of the reaction of a tertiary containing ester reagent, such as *tert*-butyl bromoacetate, with the polymer and then treating the tertiary ester of the polymer with a strong base, such as an alkali metal hydroxide, in aqueous solution, to form a carboxylate salt of the polymer.

The method may further comprise the step of treating the carboxylate salt of the polymer with an inorganic acid in aqueous solution, to convert the carboxylate salt to a carboxylic acid, thereby forming a carboxylic acid-functionalized polymer. The carboxylic acid-functionalized polymer can then be extracted from the aqueous solution with a suitable solvent, preferably a chlorinated solvent (8).

5.2 Poly(hydroxy butyrate)

Polyester-based nanoparticulate drug delivery systems, including polymer-drug conjugates and amphiphilic block copolymers, represent a major class with promising outcomes, especially for those derived from poly(3-hydroxybutyrate) (PHB). The recent advances in drug delivery systems designed from the self-assembly of synthetic (co)polymers derived from PHB have been reviewed (13).

The various strategies for the synthesis of PHB conjugates, PHB/PEG and other PHB-based copolymers were summarized. Nanoparticles, micelles, microparticles, and hydrogels elaborated from these polymers and copolymers following various preparation methods, along with their exploitation in the encapsulation and release of various therapeutic agents, were detailed. Here functionalized systems were also used (13).

Multifunctional PHB nanoparticles for theranostic applications were synthesized (14). Tri-fused protein A33scFv-green fluorescent protein-PHA synthase (AG-PHA synthase), used as targeting, imaging and catalytic moiety in the presence of 3HB-CoA, led to the formation of an AG-PHB conjugate.

Liu *et al.* clicked a 8-arm star PEG with amorphous PHB segments using the Huisgen's reaction (15), a 1,3-dipolar cycloaddition (16). This strategy was based on the complementary functionalization of PHB and PEG with alkyne and azide, respectively. Anionic ring-opening polymerization was initiated by sodium adamantane carboxylate and terminated by propargyl bromide, thereby giving α-adamantyl, ω-alkynyl functionalized PHB. So, these copolymers were used for the formation of adamantyl functionalized nanoparticles with a PEG core, and further supramolecularly assembled with heptakis(2,6-di-*o*-methyl)-β-cyclodextrin (DM-β-CD), and taking advantage of the host-guest interaction with the adamantyl chain-end functions. Heptakis(2,6-di-*o*-methyl)-β-cyclodextrin is shown in Figure 5.1.

Also, the synthesis of a peptide dendrimer-PHB-b-PEG-b-PHB conjugate has been reported (17). The two-step synthesis of these dendrimers was performed from the third generation poly(*l*-lysine) dendrimer end-functionalized with an arginine-6-oligomer. The cytotoxicity of the synthesized copolymers is evaluated with a bacterial luminescence test and protozoan assay, which showed that the obtained copolymers are not cytotoxic. Furthermore, the poly(*l*-lysine) dendrimer end-functionalized with arginine-6-oligomer was covalently bonded to a copolymer using the 1,1′-carbonyldiimidazole/4-(dimethylamino)pyridine coupling method. These compounds are shown in Figure 5.2.

5.3 Poly(glycerol)

Poly(glycerol) is an alternative for PEG and poly(propylene glycol) with regard to solubility and biological behavior.

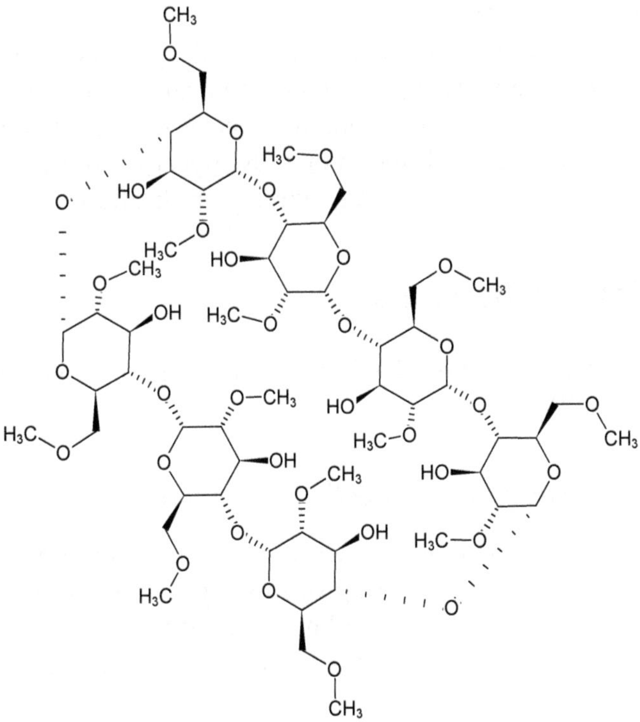

Figure 5.1 Heptakis(2,6-di-*o*-methyl)-*β*-cyclodextrin.

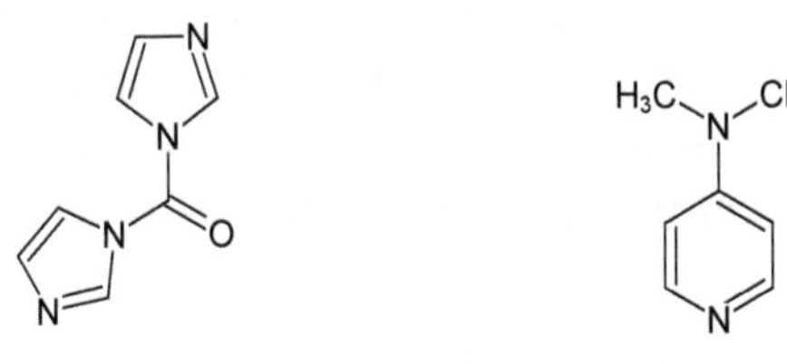

1,1′-Carbonyldiimidazole   4-(Dimethylamino)pyridine

Figure 5.2 Compounds for a coupling method (17).

Such polymers can be produced by anionic (18) or cationic (19) ring-opening polymerization of glycidol, which results in branched polymer architectures.

The main advantage of poly(glycerol) is the presence of hydroxyl groups on the main chain, which can be utilized to introduce additional functionalities to the polymer backbone (20). Examples of several dendritic scaffolds based on poly(glycerol) have been extensively studied for potential drug delivery applications.

5.4 Poly(carbonate)s

Poly(carbonate)s have received much attention for protein/peptide conjugation (21).

5.4.1 *Pentafluoro-Containing Poly(carbonate)s*

It was demonstrated that the synthesis of functional poly(carbonate)s (PCs) is possible by ring-opening polymerization starting with the monomer pentafluorophenyl 5-methyl-2-oxo-1,3-dioxane-5-carboxylate (22). The synthesis of the monomer runs as shown in Figure 5.3.

The active pentafluorophenyl esters enable a substitution with suitable nucleophiles such as alcohols and amines and the functionalization with other active groups.

The versatility of this approach was demonstrated by the preparation of numerous functional PCs, which are of particular relevance to polymer-drug conjugates, including PCs with PEG (21), hydroxyl-containing side chains (23), and zwitterionic side chains (24).

Other functional nanocarriers for biomedical applications have been extensively reviewed that highlight these PC-based degradable alternatives to PEG with minimal toxicity (23,25–27).

5.4.2 *Disulfide Five-Membered Ring Poly(carbonate)s*

A carbonate polymer was developed that contains a functional group of disulfide five-membered ring in the side chain (28).

The polymer can be prepared from a cyclic carbonate monomer containing a disulfide five-membered ring functional group through

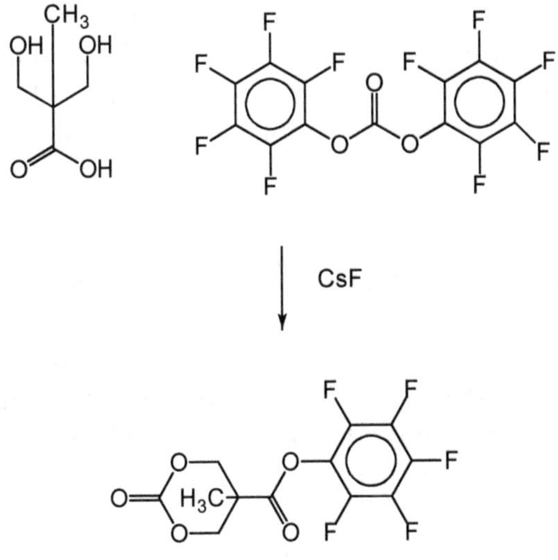

Figure 5.3 Synthesis of pentafluorophenyl 5-methyl-2-oxo-1,3-dioxane-5-carboxylate.

the activity controllable ring-opening polymerization. For the polymer, molecular weight is controlled and molecular weight distribution is narrowed and does not require the protection and deprotection procedures. Some disulfide five-membered chemicals that can be used for synthesis are shown in Figure 5.4.

1,2-Dithiolane-4,4-dimethanol 7,9-Dioxa-2,3-dithiaspiro[4.5]decan-8-one

Figure 5.4 Disulfide chemicals (28).

The polymer prepared from the carbonate monomer through the ring-opening polymerization has biodegradability and can be used for controlling drug release systems. It can also be used to prepare a tumor-targeted nano-drug carrier which is sensitive to reduction and can perform reversible crosslinking and can support long circulation in the body, and, in high concentration of cancers cells can rapidly release crosslinking in the cancer cells to release drugs to kill cancer cells with high efficiency and specificity. The biodegradable polymer has a good application value in tissue engineering and biochip coating (28).

5.5 Poly(ethylene glycol) Derivates

The conjugation of drugs to polymers is a well-established technique to improve the properties of therapeutically active substances (1). Cytostatic agents that have been mainly used for preparing polymer-drug conjugates are doxorubicin, camptothecin, paclitaxel, and platinum complexes, c.f. Figure 5.5.

The most well-known example of conjugation is the attachment of PEG, also addressed as PEGylation. PEGylation was first described in 1977 and reported for albumin and catalase modification (29).

PEGylation defines the modification of a protein, peptide or non-peptide molecule by the linking of one or more PEG chains. This polymer is nontoxic, non-immunogenic, non-antigenic, highly soluble in water and FDA approved (30). In drug delivery, PEGylation can be subdivided into three categories (1):

1. Attachment of PEG to proteins,
2. Attachment to small drug molecules, and
3. Attachment to other polymers. This is often related to the formation of nanocarriers.

The functionalization with PEG introduces several advantageous properties in comparison to the native protein. PEGylated proteins show an improved pharmacokinetic and pharmacodynamic, which is related to increased water solubility, increased stability, shielding from metabolic enzymes, reduced immunogenicity and retarded renal clearance (due to stealth behavior) leading to elongated blood circulation time (30).

The attachment of PEG chains transfers some of the beneficial properties of this polymer to pharmacological active compounds, which can improve the accessibility of drugs in biological systems (31–33).

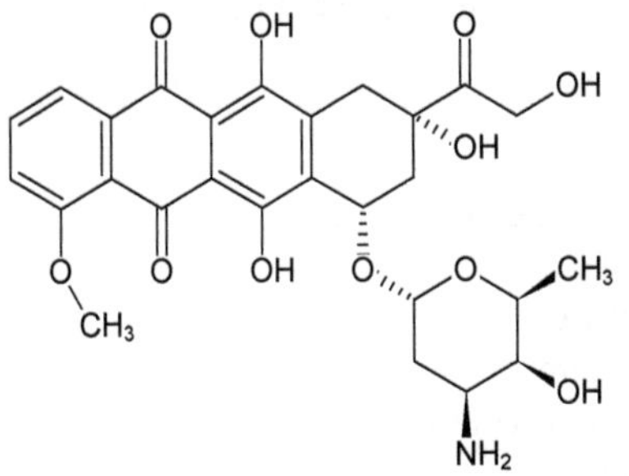

Doxorubicin

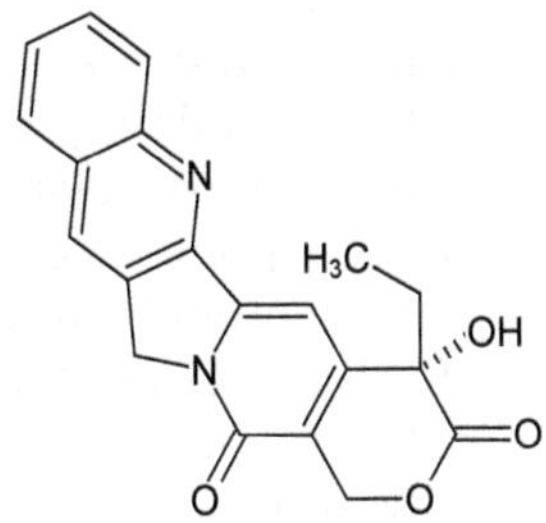

Camptothecin

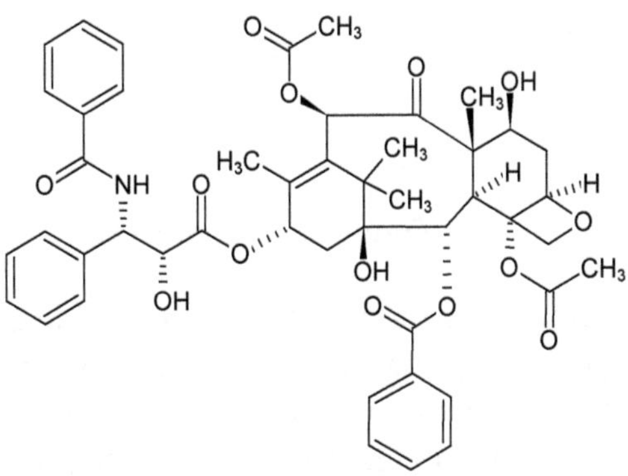

Paclitaxel

Figure 5.5 Cytostatic agents.

The most popular PEG derivatives are summarized in Table 5.2.

Table 5.2 Poly(ethylene glycol) derivatives (1).

Derivative
PEG-aldehyde (also in the form of more stable acetals)
PEG-tresyl or tosyl
PEG-dichlorotriazine or chlorotriazine, c.f. Figure 5.6
PEG-epoxide
PEG-carboxylates
PEG-succinimidyl succinate
PEG-amino acid-succinimidyl ester
PEG-peptide-succinimidyl ester
PEG-*p*-nitrophenyl carbonate
PEG-benzotriazolyl carbonate
PEG-2,3,5-trichlorophenyl carbonate
PEG-succinimidyl carbonate

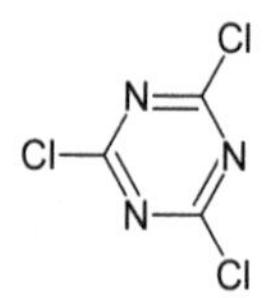

Figure 5.6 Chlorotriazine.

The specific conjugation of PEG to the amide group of glutamines or to the hydroxyl group of serines and threonines is only possible under mild conditions with the use of enzymes. There are several naturally occurring enzymes that recognize glutamine as substrate, namely specific or nonspecific transglutaminases.

It has been found that glutamine in proteins can be the substrate of the transglutaminase enzymes if an amino PEG is used as the nucleophilic donor (34). Through a transglutamination reaction the enzyme links PEG to the protein at the level of the glutamine residue.

A problem for the use of PEG is related to the excretion from the human body. As in the case of other polymers, PEGs are usually excreted in urine or feces. However, polymers with high molecular weights can accumulate in the liver, leading to a macromolecular syndrome. It is not easy to extrapolate the kidney excretion limit of

PEG by looking only at the kidney clearance threshold of protein, because other factors play an important role.

For example, the high water coordination of the polymer increases the hydrodynamic volume of PEG up to 3 to 5 times that of a globular protein with the same molecular weight, thus decreasing the polymer kidney clearance threshold and the linear and flexible structure of PEG chains that help the polymer cross the glomerular membranes by *snake-like* movements.

Nevertheless, several classes of protein drugs, such as enzymes, cytokines and antibodies, were found to be significantly improved by PEGylation (35),

A general problem encountered in PEGylation of small drugs is the low loading of this polymer, which possesses only one or two hydroxyl terminal groups that can be activated. To overcome this limitation, the construction of dendrimeric structures were proposed at the level of the polymer termini, obtaining PEG dendrons, also known as forked PEGs. The implementation of the number of active groups is reached by using branching molecules, such as bicarboxylic amino acid, which allow the increase of the drug-polymer loading (36).

5.6 Nanosized Drug Delivery Systems

In order to maximize the site-selectivity and the therapeutic efficacy, drug delivery systems should be functionalized with ligands which can specifically recognize and bind targets expressed by hepatocellular carcinoma, namely cell membrane-associated antigens, receptors or biotransporters (37).

5.7 Poly(ethylene imine)s

Poly(ethylene imine)s (PEIs) are characterized by the presence of amine group functionalities within the polymer backbone, which determine their chemical and physical behavior. PEI has one of the highest cationic charge densities of all organic macromolecules (38).

Actually, every third atom of PEI is a protonable amino nitrogen atom, which makes the polymeric network an effective *proton sponge* at virtually any pH. Luciferase reporter gene transfer with this polycation into a variety of cell lines and primary cells gave results comparable to, or even better than, lipopolyamines. The cytotoxicity is low and observed only at concentrations well above those required for optimal transfection.

This high charge density enables the interaction with the phosphate groups of genetic material, leading to the formation of toroidal complexes that are readily endocytosed by cells (39,40).

A branched PEI can be synthesized by the ring-opening polymerization of unsubstituted aziridine (41). This leads to an uncontrolled branching and chains with primary, secondary, and tertiary amine groups.

The synthesis of linear PEI was first described in the 1960s (42,43). Here, the hydrolysis of poly(2-alkyl-2-oxazoline)s was used for synthesis. 2-Aryl-2-oxazolines and 2-alkyl-2-oxazolines were polymerized to get poly(*N*-aroyl)aziridines and poly(*N*-acyl)aziridines, respectively, in the presence of boron trifluoride. The so-obtained polymers were glassy, light yellow resins with molecular weights ranging from 3500 *Dalton* to 7500 *Dalton*, which is 35 to 50 oxazoline units per chain.

The hydrolysis can be performed under acidic or basic conditions (43–47). Products are obtained that are hydrolyzed up to 97%.

5.8 Poly(amino acid)s

Almost all types of amino acids can be applied for polymerization. Therefore, a large variety of different functionalities, such as ionic or stimuli-responsive side groups can be formed, as well as complex superstructures, including micelles and gels (48–52).

Side reactions in the polymerization reaction can be suppressed using ammonium chloride-functionalized macroinitiators (53,54) or an organosilicon (55).

The functionalization of polypeptides has gained increasing attention in recent years. The introduction of various functional groups or stimuli-sensitive moieties to the side chains of the polypeptides render them particularly appealing for the design and development of multifunctional active biomaterials.

The functionalization is mainly achieved by two approaches (1):

1. A one-step ring-opening polymerization of *N*-substituted carboxyanhydride monomers already containing the desired functional moieties, or
2. The postpolymerization modification of polypeptides.

When unnatural amino acid derivatives are used as monomers, a versatility of polymer structures can be obtained with molecular and physical properties far removed from those of proteins (1). Trifunctional amino acids, such as glutamic acid, lysine and aspartic acid, are often used to achieve a structural viability within the polymers. These compounds are shown in Figure 5.7.

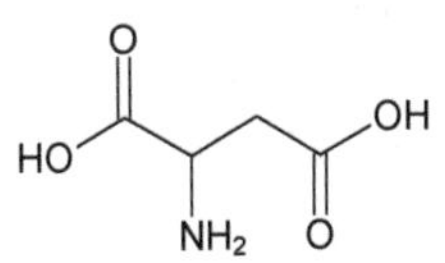

Aspartic acid

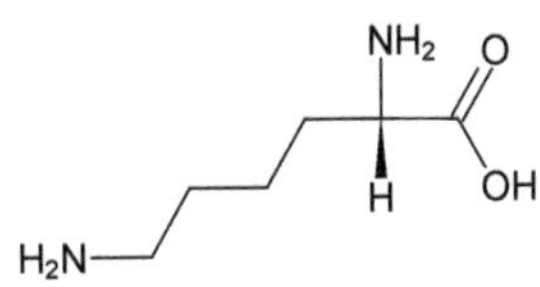

Lysine

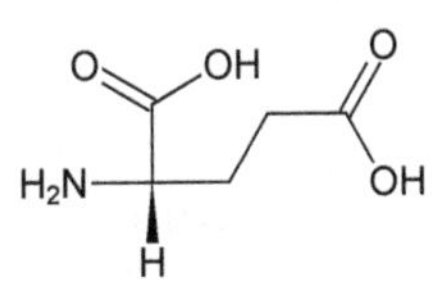

Glutamic acid

Figure 5.7 Trifunctional amino acids.

5.8.1 *Diamino Diesters*

Recent trends in biodegradable polymers indicate significant developments in terms of novel design strategies and engineering to provide advanced polymers with comparably good performance (56). Various classes of amino acid-based biodegradable polymers with a wide range of material properties, and suitable for numerous biomedical applications, were designed on the basis of diamino diester monomers. However, the scope of the application of AABB polymers could substantially be expanded by designing their functionalized analogues. This can be achieved by the combination of diamino diesters with functionalized diamino diesters or with other types of functional comonomers, or by synthesizing various active prepolymers with subsequent functionalization by means of polymer analogous reactions. Several examples of functionalized diamino diesters have been presented (56).

Various classes of amino acid-based biodegradable (AABB) polymers with a wide range of material properties, and suitable for numerous biomedical applications, were designed on the basis of diamino diester monomers (56).

Biodegradable polymers with great processing flexibility are the predominant scaffolding materials in tissue engineering. Synthetic biodegradable polymers with well-defined structure and without immunological concerns associated with naturally derived polymers are widely used in tissue engineering (57–59).

Biodegradable polymers can provide sustained/controlled release of drugs, growth factors, any bioactive molecules, and will be absorbed by the surrounding tissues, which means biodegradable polymers will be cleared from the body after executing their function (60).

In addition, biodegradable polymers discharge their degradation debris into the surrounding environment and these products can activate macrophages to produce cell growth factors, mediators, and other materials, thus can accelerate the wound healing process.

5.9 Poly(*N*-acrylamide)s

The structurally simplest example for poly(*N*-acrylamide)s is poly(acrylamide) itself, which is found in cosmetics or used for flocculation of dispersions. Its use in pharmaceutical applications is limited due to the severe toxicity of the monomer, which often can still be found in the polymer, although the polymer itself is not considered to be toxic (61).

5.9.1 Poly(N-(2-hydroxypropyl)methacrylamide)

Poly(*N*-(2-hydroxypropyl)methacrylamide) is a hydrophilic, chemically and hydrolytically stable, biocompatible polymer. The hydroxyl group functionalities allow the conjugation of drugs and targeting molecules (62).

The grafting of poly(*N*-(2-hydroxypropyl)methacrylamide) onto various surfaces has been studied using different methods, such as activated substrates and functionalized polymers (63,64), photo-initiated grafting of functionalized polymers (65–68), plasma-induced grafting (66, 67, 69), and E-beam irradiation (68, 70–73). The *N*-(2-hydroxypropyl)methacrylamide monomer is shown in Figure 5.8.

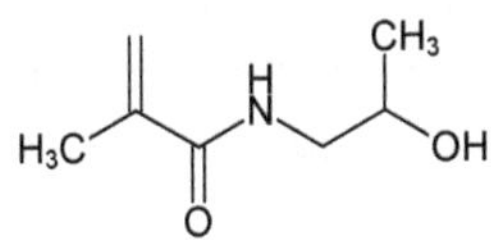

Figure 5.8 *N*-(2-Hydroxypropyl)methacrylamide.

5.10 Polyphosphates

Polyphosphates are an important class of polymers, due to their structural similarity with nucleic and teichoic acids (74). Teichoic acids are bacterial copolymers of glycerol phosphate or ribitol phosphate and carbohydrates linked via phosphodiester bonds (75).

Inorganic polyphosphates are salts or esters of polymeric oxo anions that are built on tetrahedral phosphate units. They occur as linear or branched forms, or cyclic ring structures. Commonly, these compounds are synthesized by dehydration of orthophosphate (PO_4^{3-}) at elevated temperature.

Polyphosphates may allow numerous opportunities for modifications due to the pentavalent nature of the phosphorus atoms. Examples of such modifications are the introduction of double bonds (76), triple bonds (77), hydroxy groups (78), and amine groups (79) to the side chain or to the polymer backbone.

The convenient modification of the side group of the cyclic phosphate monomers resulted in a variety of structures and functionalities. Recent advances in hyperbranched polyphosphates have been summarized (74). Hyperbranched polyphosphates with different topological structures and various functionalities were synthesized via adjusting the side group of cyclic phosphate monomers, which have shown promising biomedical applications, for example, for use as a macromolecular anticancer agent and in the construction of advanced drug delivery systems, including site-specific delivery systems, self-delivery systems, and stimuli-responsive delivery systems.

Under physiological conditions, polyphosphates can degrade into nontoxic, low molar mass species through the hydrolysis or enzymatic cleavage of the phosphate bond (1).

Polyphosphates have been shown to stabilize the protein structure by forming a surface layer of coagulated protein around meat or conserving the natural moisture of seafood.

5.11 Poly(vinyl ether)s

Poly(vinyl ether)s are polymers bearing the functional ether group in the side chain. They can be fabricated from the respective vinyl ethers via chain-growth polymerization. The functional versatility of the starting material allows the incorporation of alkyl or amine moieties.

Various synthesis routes and industrial processes have been described (80). Probably the most important poly(vinyl ether) is the amphipathic butyl-modified and amine-modified one that has shown remarkable performances in trial experiments.

5.12 Poly(*N*-vinyl amide)s

Poly(*N*-vinyl amide)s belong to a polymer class that comprises amide functionalities in the side chain. Compared to poly(*N*-acryl amide)s, the amide group is linked to the polymer backbone through the nitrogen atom. The most prominent representative is poly(*N*-vinyl-2-pyrrolidone) (PVP).

PVP is a water-soluble, nonionic, biocompatible and stable polymer, which is not metabolized by the organism (81). Linear multifunctional PVP chains bearing carboxylate ($-COO^-$) and/or sulfonate ($-SO^{3-}$) groups have been prepared using a standard radical copolymerization reaction of homologous vinylpyrrolidone and the following vinylpyrrolidone derivatives: 3-carboxyvinylpyrrolidone and 3-sulfoalkylvinylpyrrolidone. Functional PVP networks with a homogeneous crosslinking density could also be prepared using a homologous water-compatible VP-R-VP crosslinking agent. The compounds are shown in Figure 5.9.

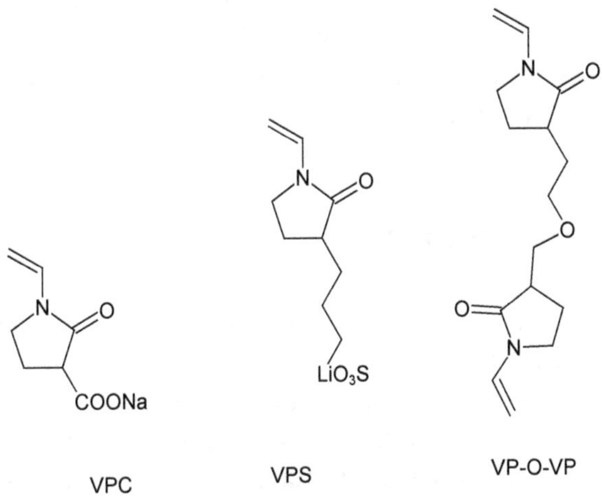

Figure 5.9 Vinylpyrrolidone derivatives. VPC: 3-Carboxyvinylpyrrolidone, VPS: 3-Sulfoalkylvinylpyrrolidone, VP-O-VP: Crosslinking agent.

The synthesis of the crosslinking agent runs as follows (81):

Preparation 5–1: A freshly distilled solution of vinylpyrrolidone in anhydrous tetrahydrofuran (THF) was added dropwise to a commercial solution of lithium diisopropylamide (2.0 *M*, in THF, hexane, and ethylbenzene) in inert atmosphere and at −78°C and stirred for 2 *h*. Then, 2-bromoethyl ether was added dropwise. The resulting solution was magnetically stirred for 2 *h* at −78°C, and then at room temperature overnight. At the end of the reaction, monitored by thin-layer chromatography, the solution was hydrolyzed in CH_2Cl_2:H_2 = 2:1, total amount 300 *ml*.

The aqueous layer was extracted with CH_2Cl_2 and the organic layers were combined and dried over sodium sulfate. The solvent was evaporated at low pressure.

This homologous approach for the preparation of functionalized PVP systems was compared theoretically with the preparation of vinylpyrrolidone with commercial acrylic comonomers bearing similar functionalities. For example, the tentative functionalization of PVP can be achieved by the copolymerization with methacrylic compounds. Methacrylic acid effects the incorporation of carboxyl groups in the polymeric backbone.

All the linear chains and networks were tested under basic *in vitro* cell cultures. It was found that all copolymers in the concentration range of 1–1000 $\mu g\, ml^{-1}$ were highly biocompatible, with recovered deoxyribonucleic acid quantities similar to those recovered from the positive PVP controls (81).

5.13 Poly(allylamine)

Poly(allylamine) is a cationic polymer that can be obtained from the radical or cationic polymerization of allylamine. While the backbone contains no nitrogen, as described also for poly(ethylene imine), the polymer side chain contains primary amine groups which can be converted into secondary or tertiary amine functionalities. This converts poly(allylamine) into a highly promising gene delivery agent (82).

Poly(allylamine hydrochloride), a water-soluble biocompatible synthetic polymer, has been proposed as a non-viral transfection agent (83). However, poly(allylamine hydrochloride) alone is toxic and has a relatively low transfection efficiency, which makes it a rather unlikely candidate as a gene carrier in its native form (84). So, the primary amino groups of poly(allylamine) were substituted with imidazolyl functions, which are presumed to enhance endosomal release, and thus enhance its gene delivery efficiency and eliminate the requirement of external lysosomotropic agents.

Also, other attempts were made to improve transfection efficiency and reduce toxicity of a poly(allylamine hydrochloride) polymer by various chemical modifications such as: glycolylation (85), guanidylation (86,87), thiolation (88,89) of the amino groups or substitution by imidazole groups (90).

In the case of guanidylation, poly(allylamine) was reacted with 6-(*N*,*N*,*N*′,*N*′-tetramethylguanidinium chloride) hexanoic acid to prepare tetramethylguanidinium poly(allylamine) polymers, which can be used as vectors for gene transfection (87).

Poly(allylamine hydrochloride) in combination with an anionic polyelectrolyte (e.g., poly(styrene sulfonate)) can be used to form layer-by-layer adsorbed films (91).

5.14 Poly((meth)acrylate)s

Poly(acrylate)s and poly(methacrylate)s are synthetic polymers made from acrylic and methacrylic acids or their esters, respectively. The properties of poly(meth)acrylates can be tuned by varying the molecular structure of the ester side chain. These modifications enable access to polymers spanning the whole range from water to oil soluble or from brittle to elastic.

The FDA has approved the safety of poly(meth)acrylates in several pharmaceutical applications. The backbones of poly(acrylic acid) and poly(methacrylic acid) are not biodegradable. However, they exhibit a low toxicity and excellent biocompatibility (92).

An advantage of poly(acrylic acid) is the convenient modification of the carboxyl side chains with alcohols or amines to introduce additional functionalities. Therefore, drugs and/or bioactive molecules can easily be attached to the poly(acrylic acid) backbone in accordance with the concept of Ringsdorf (93, 94).

In 1975, Ringsdorf proposed a macromolecular pro-drug model consisting of multiple drug molecules attached to a macromolecule, namely a multifunctional polymer, through bonds that can be cleaved and which release the drug in the disease site (37).

An important modification of poly(methyl methacrylate) is poly-(2-dimethylaminoethyl methacrylate). The monomer is shown in Figure 5.10.

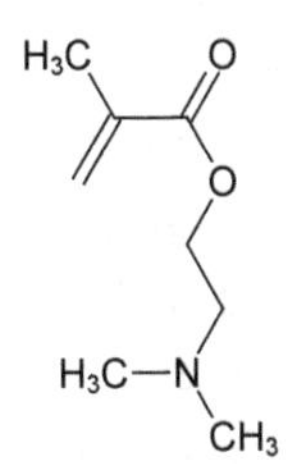

Figure 5.10 2-Dimethylaminoethyl methacrylate.

This material contains an ionizable tertiary amine group. In water it may be protonated and becomes a weak cationic polyelectrolyte depending on the pH value. The pK_a value of this polymer depends on the composition, as well as the molar mass.

Poly(2-dimethylaminoethyl methacrylate) exhibits a lower critical solution temperature behavior in water depending on the molar mass of the polymer and the pH value of the solution (95).

However, the positive charge on the polymer causes these materials to be quite cytotoxic, which, in combination with the lack of biodegradability, limits its use in pharmaceutical applications (96).

5.15 Poly(acrylonitrile)s

Poly(acrylonitrile)s (PANs) are synthetic, semicrystalline organic polymers consisting of nitrile units attached to the carbon backbone. The relatively poor biocompatibility of conventional PAN can be improved by bulk or surface modification (97).

The active nitrile groups in PAN can be converted to other functional groups like carboxyls via hydrolysis (98) and amines via reduction (99), which subsequently facilitate further modifications.

Residues of pharmaceuticals are potentially hazardous contaminants to aquatic life and humans. Pharmaceutical residues have been detected in Malaysian tropical wastewaters (100). A challenge concerning this issue is the development of enrichment techniques able to extract polar pharmaceutical residues, since these compounds are widely found in aqueous samples.

An acrylonitrile-divinylbenzene-80-vinylbenzylchloride porous terpolymer material was prepared via a precipitation polymerization method. The porous terpolymer containing chlorine pendant groups was hypercrosslinked via a Friedel-Crafts reaction to develop 3D network structure within the terpolymer chains.

The hypercrosslinked porous material was then chemically modified with ethylenediamine to develop active functional groups (diamine moisties) along terpolymer chains.

Their high specific surface area and the polar character (arising from acrylic residues) of these materials suggest their use as potential materials to extract pharmaceutical residues (100).

5.16 Antibacterial Agents

Amino-functionalized poly(glycidyl methacrylate)-based microspheres were prepared via soap-free emulsion polymerization (101). Glycidyl methacrylate is shown in Figure 5.11.

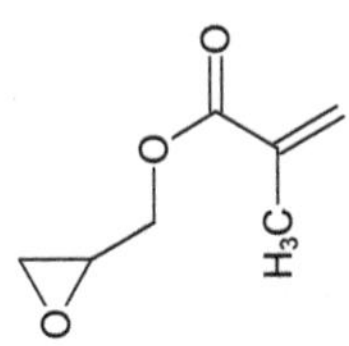

Figure 5.11 Glycidyl methacrylate.

These materials served as polymeric matrices for silver nanoparticles loading and were applied as antibacterial agent. The amino and hydroxyl enriched microspheres served as both matrix for Ag^+ absorbance through coordination bond and reducing agent under hydrothermal condition in a high-pressure steam sterilization.

The effects of the chemical structure of the microspheres on the morphology of silver nanoparticles/polymer composites were investigated (101). The composite spheres with ultrafine silver nanoparticles up to 26% were well characterized and their antibacterial activity was studied in detail. It was found that the silver nanoparticle loading in microspheres showed comparable antibacterial activities in terms of Ag with the minimum inhibitory concentration of low to 20 $mg\,L^{-1}$ and 20 $mg\,L^{-1}$ (equal to 5.2 $mg\,L^{-1}$, 5.2 $mg\,L^{-1}$ Ag) for *Escherichia coli* and *Staphylococcus aureus*, respectively.

Thus, the synthesized composite spheres are promising alternative antibacterial agents for industrial and biomedical applications (101).

5.17 Clenbuterol Analysis

Clenbuterol is a β_2 adrenergic agonist, which can be used to expand the bronchi and miscarriage and as clinical medicine for the treatment of asthma (102). It can also promote animal growth, more muscle tissue and reduce animal fat, so it is illegally used as growth promoter in livestock and poultry production to result in leaner meat with a higher muscle-to-fat ratio. When a clenbuterol nutrient redistribution agent was eaten by animals, it could not be completely metabolized and could accumulate in animal tissue and fluid.

After consuming the products of livestock or poultry fed with feedstock containing clenbuterol, people will get food poisoning and display symptoms such as muscle tremors, tachycardia, palpitations, and dizziness, which in some cases may even result in death (102).

For the rapid and robust detection of both parent clenbuterol and its metabolites in swine urine samples, a novel quartz crystal microbalance sensor array for clenbuterol detection based on molecularly imprinted polymers was developed (102).

At first, clenbuterol and the structural analogues of its metabolites, 4-aminohippuric acid and 4-hydroxymandelic acid, were chosen as molecular templates. These compounds are shown in Figure 5.12

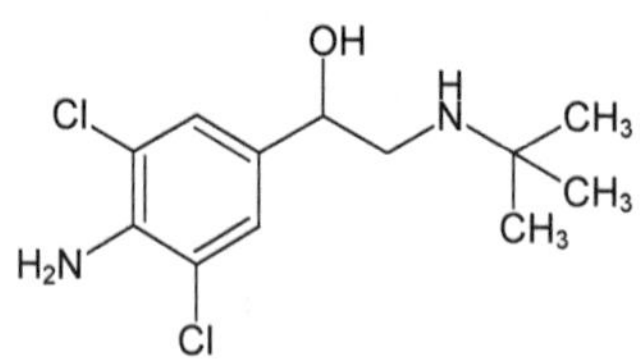

Clenbuterol

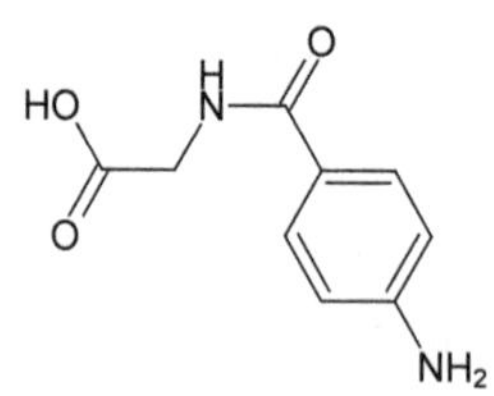

4-Aminohippuric acid

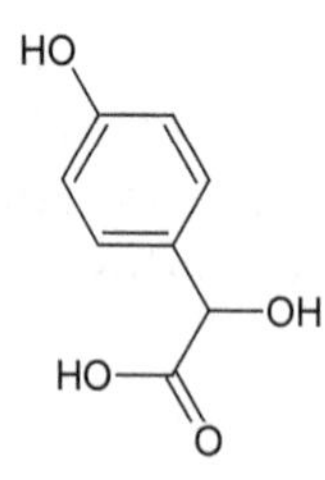

4-Hydroxymandelic acid

Figure 5.12 Clenbuterol and the structural analogues.

By computational molecular modeling, the optimum ratio between the functional monomer and molecular template was selected. A surface imprinting method was applied to modify the surface of the quartz crystal microbalance electrode to graft a thin molecularly imprinted polymer film.

The study demonstrated that the developed method could be applied to detect whether the livestock was fed with clenbuterol nutrient redistribution agent by checking the urine samples (102).

References

1. C. Englert, J.C. Brendel, T.C. Majdanski, T. Yildirim, S. Schubert, M. Gottschaldt, N. Windhab, and U.S. Schubert, *Progress in Polymer Science*, 2018. in press.
2. B. Obermeier, F. Wurm, C. Mangold, and H. Frey, *Angewandte Chemie International Edition*, Vol. 50, p. 7988, 2011.
3. C. Mangold, F. Wurm, and H. Frey, *Polym. Chem.*, Vol. 3, p. 1714, 2012.
4. M.J. Barthel, T. Rudolph, S. Crotty, F.H. Schacher, and U.S. Schubert, *Journal of Polymer Science Part A: Polymer Chemistry*, Vol. 50, p. 4958, 2012.
5. A.L. Klibanov, K. Maruyama, V.P. Torchilin, and L. Huang, *FEBS Letters*, Vol. 268, p. 235, 1990.
6. G. Blume and G. Cevc, *BBA - Biomembr.*, Vol. 1029, p. 91, 1990.
7. S. Schöttler, G. Becker, S. Winzen, T. Steinbach, K. Mohr, K. Landfester, V. Mailänder, and F.R. Wurm, *Nature Nanotechnology*, Vol. 11, p. 372, February 2016.
8. J.K. Bardman and W.T. Brown, Coating with improved hiding, compositions prepared therewith, and processes for the preparation thereof, US Patent 7 081 488, assigned to Rohm and Haas Company (Philadelphia, PA), July 25, 2006.
9. F.M. Veronese, P. Caliceti, A. Pastorino, O. Schiavon, L. Sartore, L. Banci, and L.M. Scolaro, *Journal of Controlled Release*, Vol. 10, p. 145, 1989.
10. J.J. Krepinsky, S.P. Douglas, and D.M. Whitfield, Polymer-supported solution synthesis of oligosaccharides, US Patent 5 278 303, assigned to University of Toronto Innovations Foundation (Ontario, CA), January 11, 1994.
11. A.J. Martinez and R.B. Greenwald, Method of preparing polyalkylene oxide carboxylic acids, US Patent 5 681 567, assigned to Enzon, Inc. (Piscataway, NJ), October 28, 1997.
12. M. Donbrow, *Stability of the polyoxyethylene chain* in M.J. Schick, ed., *Nonionic Surfactants Physical Chemistry*, Vol. 23 of *Surfactant Science Series*, chapter 18, pp. 1011–1072. Marcel Dekker, New York, 1987.

13. G. Barouti, C.G. Jaffredo, and S.M. Guillaume, *Progress in Polymer Science*, Vol. 73, p. 1, 2017.
14. H.-S. Kwon, S.-G. Jung, H.-Y. Kim, S.A. Parker, C.A. Batt, and Y.-R. Kim, *J. Mater. Chem. B*, Vol. 2, p. 3965, 2014.
15. Wikipedia contributors, Azide-alkyne Huisgen cycloaddition — Wikipedia, the free encyclopedia, https://en.wikipedia.org/w/index.php?title=Azide-alkyne_Huisgen_cycloaddition&oldid=852108699, 2018. [Online; accessed 30-August-2018].
16. J.-L. Zhu, K.L. Liu, Z. Zhang, X.-Z. Zhang, and J. Li, *Chem. Commun.*, Vol. 47, p. 12849, 2011.
17. E. Oledzka, P. Sliwerska, M. Sobczak, B. Kraska, W. Kamysz, G. Nalecz-Jawecki, and W. Kolodziejski, *Macromolecular Chemistry and Physics*, Vol. 216, p. 1365, 2015.
18. M. Schömer, C. Schüll, and H. Frey, *Journal of Polymer Science Part A: Polymer Chemistry*, Vol. 51, p. 995, 2012.
19. P. Kubisa, *Journal of Polymer Science Part A: Polymer Chemistry*, Vol. 41, p. 457, 2002.
20. M.A. Quadir and R. Haag, *Journal of Control Release*, Vol. 161, p. 484, 2012.
21. S. Tempelaar, L. Mespouille, O. Coulembier, P. Dubois, and A.P. Dove, *Chemical Society Reviews*, Vol. 42, p. 1312, 2013.
22. D.P. Sanders, K. Fukushima, D.J. Coady, A. Nelson, M. Fujiwara, M. Yasumoto, and J.L. Hedrick, *Journal of the American Chemical Society*, Vol. 132, p. 14724, 2010.
23. A.C. Engler, X. Ke, S. Gao, J.M.W. Chan, D.J. Coady, R.J. Ono, R. Lubbers, A. Nelson, Y.Y. Yang, and J.L. Hedrick, *Macromolecules*, Vol. 48, p. 1673, 2015.
24. J.M.W. Chan, X. Ke, H. Sardon, A.C. Engler, Y.Y. Yang, and J.L. Hedrick, *Chem. Sci.*, Vol. 5, p. 3294, 2014.
25. Z.Y. Ong, K. Fukushima, D.J. Coady, Y.-Y. Yang, P.L.R. Ee, and J.L. Hedrick, *Journal of Controlled Release*, Vol. 152, p. 120, 2011. Symposium on Innovative Polymers for Controlled Delivery.
26. J. Feng, R.-X. Zhuo, and X.-Z. Zhang, *Progress in Polymer Science*, Vol. 37, p. 211, 2012. 37/2 Topical Issue on Biomaterials.
27. W. Chen, F. Meng, R. Cheng, C. Deng, J. Feijen, and Z. Zhong, *Journal of Controlled Release*, Vol. 190, p. 398, 2014. 30th Anniversary Special Issue.
28. F. Meng, Y. Zou, Z. Zhong, and J. Yuan, Carbonate polymer containing a functional group of disulfide five-membered ring in the side chain and application thereof, US Patent 10 072 122, assigned to Brightgene Bio-Medical Technology (Suzhou) Co., Ltd. (Suzhou, CN), September 11, 2018.
29. A. Abuchowski, T. Van Es, N.C. Palczuk, and F.F. Davis, *Journal of Biological Chemistry*, Vol. 252, p. 3578, 1977.

30. F.M. Veronese and G. Pasut, *Drug Discovery Today*, Vol. 10, p. 1451, 2005.
31. K. Knop, R. Hoogenboom, D. Fischer, and U.S. Schubert, *Angewandte Chemie International Edition*, Vol. 49, p. 6288, 2010.
32. A. Grigoletto, K. Maso, A. Mero, A. Rosato, O. Schiavon, and G. Pasut, *Journal of Drug Delivery Science and Technology*, Vol. 32, p. 132, 2016.
33. P. Mishra, B. Nayak, and R.K. Dey, *Asian Journal of Pharmaceutical Sciences*, Vol. 11, p. 337 , 2016.
34. H. Sato, *Advanced Drug Delivery Reviews*, Vol. 54, p. 487, 2002.
35. G. Pasut, A. Guiotto, and F.M. Veronese, *Expert Opinion on Therapeutic Patents*, Vol. 14, p. 859, 2004.
36. Y.H. Choe, C.D. Conover, D. Wu, M. Royzen, Y. Gervacio, V. Borowski, M. Mehlig, and R.B. Greenwald, *Journal of Controlled Release*, Vol. 79, p. 55 , 2002.
37. C. Turato, A. Balasso, V. Carloni, C. Tiribelli, F. Mastrotto, A. Mazzocca, and P. Pontisso, *Journal of Controlled Release*, Vol. 268, p. 184, 2017.
38. O. Boussif, F. Lezoualc'h, M.A. Zanta, M.D. Mergny, D. Scherman, B. Demeneix, and J.P. Behr, *Proceedings of the National Academy of Sciences*, Vol. 92, p. 7297, 1995.
39. W.T. Godbey, K.K. Wu, and A.G. Mikos, *Journal of Controlled Release*, Vol. 60, p. 149, 1999.
40. W.T. Godbey, M.A. Barry, P. Saggau, K.K. Wu, and A.G. Mikos, *Journal of Biomedical Materials Research*, Vol. 51, p. 321, 2000.
41. G.D. Jones, A. Langsjoen, S.M.M.C. Neumann, and J. Zomlefer, *The Journal of Organic Chemistry*, Vol. 09, p. 125, 1944.
42. T. Kagiya, S. Narisawa, T. Maeda, and K. Fukui, *Journal of Polymer Science Part B: Polymer Letters*, Vol. 4, p. 441, 1966.
43. D.A. Tomalia and D.P. Sheetz, *Journal of Polymer Science Part A-1: Polymer Chemistry*, Vol. 4, p. 2253, 1966.
44. R. Tanaka, I. Ueoka, Y. Takaki, K. Kataoka, and S. Saito, *Macromolecules*, Vol. 16, p. 849, 1983.
45. L. Tauhardt, K. Kempe, K. Knop, E. Altuntas, M. Jäger, S. Schubert, D. Fischer, and U.S. Schubert, *Macromolecular Chemistry and Physics*, Vol. 212, p. 1918, 2011.
46. T. Saegusa, H. Ikeda, and H. Fujii, *Macromolecules*, Vol. 5, p. 108, 1972.
47. T. Saegusa, S. Kobayashi, and A. Yamada, *Macromolecules*, Vol. 8, p. 390, 1975.
48. A. Lavasanifar, J. Samuel, and G.S. Kwon, *Advanced Drug Delivery Reviews*, Vol. 54, p. 169, 2002.
49. C. Li, *Advanced Drug Delivery Reviews*, Vol. 54, p. 695, 2002.
50. Y. Bae and K. Kataoka, *Advanced Drug Delivery Reviews*, Vol. 61, p. 768, 2009.

51. S.F.M. van Dongen, H.-P.M. de Hoog, R.J.R.W. Peters, M. Nallani, R.J.M. Nolte, and J.C.M. van Hest, *Chemical Reviews*, Vol. 109, p. 6212, 2009.
52. S. Hehir and N.R. Cameron, *Polymer International*, Vol. 63, p. 943, 2014.
53. I. Dimitrov and H. Schlaad, *Chem. Commun.*, pp. 2944–2945, 2003.
54. J.-F. Lutz, D. Schütt, and S. Kubowicz, *Macromolecular Rapid Communications*, Vol. 26, p. 23, 2004.
55. H. Lu and J. Cheng, *Journal of the American Chemical Society*, Vol. 129, p. 14114, 2007.
56. N. Zavradashvili, G. Jokhadze, M. Gverdtsiteli, D. Tugushi, and R. Katsarava, *Res. Rev. Polym*, Vol. 8, p. 105, 2017.
57. D.S. Kohane and R. Langer, *Pediatric Research*, Vol. 63, p. 487, May 2008.
58. H. Pinnock, M. Thomas, I. Tsiligianni, K. Lisspers, A. Østrem, B. Ställberg, O. Yusuf, D. Ryan, J. Buffels, J.W.L. Cals, N.H. Chavannes, S.H. Henrichsen, A. Langhammer, E. Latysheva, C. Lionis, J. Litt, T. van der Molen, N. Zwar, and S. Williams, *Primary Care Respiratory Journal*, Vol. 19, p. S1, May 2010.
59. S. Sowmya, P.T.S. Kumar, K.P. Chennazhi, S.V. Nair, H. Tamura, and J. Rangasamy, *Artificial Organs*, Vol. 25, p. 1, 2011.
60. M. Amiji and K. Park, *Journal of Biomaterials Science, Polymer Edition*, Vol. 4, p. 217, 1993.
61. D.J. King and R.R. Noss, *Reviews on environmental health*, Vol. 8, p. 3, 1989.
62. Z.M. Rzaev, S. Dinçer, and E. Piskin, *Progress in Polymer Science*, Vol. 32, p. 534, 2007.
63. H.A. von Recum, S.W. Kim, A. Kikuchi, M. Okuhara, Y. Sakurai, and T. Okano, *Journal of Biomedical Materials Research*, Vol. 40, p. 631, 1998.
64. T. Peng and Y.-L. Cheng, *Journal of Applied Polymer Science*, Vol. 70, p. 2133, 1998.
65. H. Kanazawa, Y. Kashiwase, K. Yamamoto, Y. Matsushima, A. Kikuchi, Y. Sakurai, and T. Okano, *Analytical Chemistry*, Vol. 69, p. 823, 1997.
66. S. Åkerman, P. Viinikka, B. Svarfvar, K. Putkonen, K. Järvinen, K. Kontturi, J. Näsman, A. Urtti, and P. Paronen, *International Journal of Pharmaceutics*, Vol. 164, p. 29, 1998.
67. T. Okano, N. Yamada, H. Sakai, and Y. Sakurai, *Journal of Biomedical Materials Research*, Vol. 27, p. 1243, 1993.
68. M. Yamato, M. Okuhara, F. Karikusa, A. Kikuchi, Y. Sakurai, and T. Okano, *Journal of Biomedical Materials Research*, Vol. 44, p. 44, 1999.
69. A. Kikuchi and T. Okano, *Progress in Polymer Science*, Vol. 27, p. 1165, 2002.

70. N. Yamada, T. Okano, H. Sakai, F. Karikusa, Y. Sawasaki, and Y. Sakurai, *Die Makromolekulare Chemie, Rapid Communications*, Vol. 11, p. 571, 1990.
71. T. Okano, N. Yamada, M. Okuhara, H. Sakai, and Y. Sakurai, *Biomaterials*, Vol. 16, p. 297, 1995.
72. Y.M. Lee and J.K. Shim, *Polymer*, Vol. 38, p. 1227, 1997.
73. H. von Recum, T. Okano, and S.W. Kim, *Journal of Controlled Release*, Vol. 55, p. 121, 1998.
74. J. Liu, W. Huang, Y. Pang, and D. Yan, *Chem. Soc. Rev.*, Vol. 44, p. 3942, 2015.
75. Wikipedia contributors, Teichoic acid — Wikipedia, the free encyclopedia, https://en.wikipedia.org/w/index.php?title=Teichoic_acid&oldid=841414288, 2018. [Online; accessed 18-August-2018].
76. J.-Z. Du, T.-M. Sun, S.-Q. Weng, X.-S. Chen, and J. Wang, *Biomacromolecules*, Vol. 8, p. 3375, 2007.
77. S. Zhang, A. Li, J. Zou, L.Y. Lin, and K.L. Wooley, *ACS Macro Letters*, Vol. 1, p. 328, 2012.
78. W.-J. Song, J.-Z. Du, N.-J. Liu, S. Dou, J. Cheng, and J. Wang, *Macromolecules*, Vol. 41, p. 6935, 2008.
79. J. Wang, H.-Q. Mao, and K.W. Leong, *Journal of the American Chemical Society*, Vol. 123, p. 9480, 2001.
80. G. Schröder, *Poly(vinyl ethers)* in *Ullmann's Encyclopedia of Industrial Chemistry*. American Cancer Society, 2000.
81. M.G. Tardajos, M. Nash, Y. Rochev, H. Reinecke, C. Elvira, and A. Gallardo, *Macromolecular Chemistry and Physics*, Vol. 213, p. 529, 2012.
82. M. Wytrwal, C. Leduc, M. Sarna, C. Goncalves, M. Kepczynski, P. Midoux, M. Nowakowska, and C. Pichon, *International Journal of Pharmaceutics*, Vol. 478, p. 372, 2015.
83. Y. Zhou and Y. Li, *Biophysical Chemistry*, Vol. 107, p. 273, 2004.
84. A. Pathak, A. Aggarwal, R.K. Kurupati, S. Patnaik, A. Swami, Y. Singh, P. Kumar, S.P. Vyas, and K.C. Gupta, *Pharmaceutical Research*, Vol. 24, p. 1427, August 2007.
85. O. Boussif, T. Delair, C. Brua, L. Veron, A. Pavirani, and H.V.J. Kolbe, *Bioconjugate Chemistry*, Vol. 10, p. 877, 1999.
86. J.-H. Yu, J. Huang, H.-L. Jiang, J.-S. Quan, M.-H. Cho, and C.-S. Cho, *Journal of Applied Polymer Science*, Vol. 112, p. 926, 2009.
87. M. Mahato, G. Rana, P. Kumar, and A.K. Sharma, *Journal of Polymer Science Part A: Polymer Chemistry*, Vol. 50, p. 2344, 2012.
88. I. Bacalocostantis, V.P. Mane, M.S. Kang, A.S. Goodley, S. Muro, and P. Kofinas, *Biomacromolecules*, Vol. 13, p. 1331, 2012.
89. I. Bacalocostantis, V.P. Mane, A.S. Goodley, W.E. Bentley, S. Muro, and P. Kofinas, *Journal of Biomaterials Science, Polymer Edition*, Vol. 24, p. 912, 2013.

90. A. Pathak, S. Patnaik, and K.C. Gupta, *Biotechnology Journal*, Vol. 4, p. 1559, 2009.
91. J.M.C. Lourenço, P.A. Ribeiro, A.M. Botelho do Rego, F.M. Braz Fernandes, A.M.C. Moutinho, and M. Raposo, *Langmuir*, Vol. 20, p. 8103, 2004.
92. V.G. Kadajji and G.V. Betageri, *Polymers*, Vol. 3, p. 1972, 2011.
93. G.M. Eichenbaum, P.F. Kiser, A.V. Dobrynin, S.A. Simon, and D. Needham, *Macromolecules*, Vol. 32, p. 4867, 1999.
94. A. Bernkop-Schnürch, C. Egger, M.E. Imam, and A.H. Krauland, *Journal of Controlled Release*, Vol. 93, p. 29, 2003.
95. R. Yañez-Macias, I. Alvarez-Moises, I. Perevyazko, A. Lezov, R. Guerrero-Santos, U.S. Schubert, and C. Guerrero-Sanchez, *Macromolecular Chemistry and Physics*, Vol. 218, p. 1700065, 2017.
96. Y.-Z. You, D.S. Manickam, Q.-H. Zhou, and D. Oupický, *Journal of Controlled Release*, Vol. 122, p. 217, 2007. Proceedings of the Thirteenth International Symposium on Recent Advances in Drug Delivery Systems.
97. C. Dizman, D.O. Demirkol, S. Ates, L. Torun, S. Sakarya, S. Timur, and Y. Yagci, *Colloids and Surfaces B: Biointerfaces*, Vol. 88, p. 265, 2011.
98. M.L. Gupta, B. Gupta, W. Oppermann, and G. Hardtmann, *Journal of Applied Polymer Science*, Vol. 91, p. 3127, 2004.
99. N. Arsalani, R. Rakh, E. Ghasemi, and A.A. Entezami, Vol. 18, p. 623, 2009.
100. N.N.S. Subri, M. Jamil, S.N. Ain, M.F. Ismail, A. Manap, M. Rashidi, T. Munshi, I. Scowen, et al., Preparation of chemically modified and hypercrosslinked microspheres of poly(acrylonitrile-*co*-divinylbenzene-80-*co*-vinylbenzylchloride) as sorbents to capture pharmaceutical residues, Monash University Malaysia, 2018. Monash Science Symposium (MSS).
101. Y. Deng, J. Li, Y. Pu, Y. Chen, J. Zhao, and J. Tang, *Reactive and Functional Polymers*, Vol. 103, p. 92, 2016.
102. F. Feng, J. Zheng, P. Qin, T. Han, and D. Zhao, *Talanta*, Vol. 167, p. 94, 2017.

Index

Acronyms

Chemicals

Boldface numbers refer to Figures

General Index

Also of Interest

Check out these other books by the author published by Scrivener Publishing

3D Industrial Printing with Polymers
By Johannes Karl Fink
Published 2018. ISBN 978-1-119-55526-1

Polymer Waste Management
By Johannes Karl Fink
Published 2018. ISBN 978-1-119-53608-6

Materials, Chemicals and Methods for Dental Applications
By Johannes Karl Fink
Published 2018. ISBN 978-1-119-51031-4

Fuel Cells, Solar Panels, and Storage Devices
Materials and Methods
By Johannes Karl Fink
Published 2018. ISBN 978-1-119-48010-5

Chemicals and Methods for Conservation and Restoration
Paintings, Textiles, Fossils, Wood, Stones, Metals, and Glass
By Johannes Karl Fink
Published 2017. ISBN 978-1-119-41824-5

Additives for High Performance Applications
Chemistry and Applications
By Johannes Karl Fink
Published 2017. ISBN 978-1-119-36361-3

Metallized and Magnetic Polymers
By Johannes Karl Fink
Published 2016. ISBN: 978-1-119-24232-1

Marine, Waterborne, and Water-Resistant Polymers
Chemistry and Applications
By Johannes Karl Fink
Published 2015. ISBN 978-1-119-018486-7

The Chemistry of Printing Inks and Their Electronics and Medical Applications
By Johannes Karl Fink
Published 2015. ISBN 978-1-119-04130-6

The Chemistry of Bio-based Polymers
By Johannes Karl Fink
Published 2014. ISBN 978-1-118-83725-2

Polymeric Sensors and Actuators
By Johannes Karl Fink
Published 2012. ISBN 978-1-118-41408-8

Handbook of Engineering and Specialty Thermoplastics
Part 1: Polyolefins and Styrenics
By Johannes Karl Fink
Published 2010. ISBN 978-0-470-62483-5

Handbook of Engineering and Specialty Thermoplastics
Part 2: Water Soluble Polymers
By Johannes Karl Fink
Published 2011. ISBN 978-1-118-06275-3

A Concise Introduction to Additives for Thermoplastic Polymers
by Johannes Karl Fink.
Published 2010. ISBN 978-0-470-60955-2

Lightning Source UK Ltd.
Milton Keynes UK
UKHW020507031019
350823UK00001B/2/P

9 781119 592020